MARINE DIESEL ENGINES

Maintenance and Repair Manual

Jean-Luc Pallas

OTHER TITLES OF INTEREST

MOTORBOAT ELECTRICAL & ELECTRONICS MANUAL
John C. Payne

Following the international success of *Marine Electrical and Electronics Bible*,
Payne turns his talents from sailing boats to powerboats.

This complete guide, which covers inboard engine boats of all ages, types and sizes,
is a must for all builders, owners and operators.

Contents include: diesel engines, instrumentation and control, bow thrusters, stabilizers, A/C and
refrigeration, water and sewage systems, batteries and charging, wiring systems, corrosion,
AC power systems, generators, fishfinders, sonar, computers, charting and GPS, radar, autopilots,
GDMSS, radio frequencies and more.

"... *tells the reader how to maintain or upgrade just about every type of inboard engine vessel.*"
Soundings

MARINE ELECTRICAL & ELECTRONICS BIBLE, 3rd ed.
John C. Payne

"*A bible this really is...the clarity and attention to detail make this an ideal reference book that every
professional and serious amateur fitter should have to hand.*"
Cruising

"*All in all, this book makes an essential reference manual for both the uninitiated and the expert.*"
Yachting Monthly

"*...a concise, useful, and thoroughly practical guide.... It's a 'must-have-on-board' book.*"
Sailing & Yachting (SA)

UNDERSTANDING BOAT BATTERIES
John C. Payne

This new series features easy-to-understand yet thorough treatments of technical issues facing every
boat owner, sail or power. Each volume is fully illustrated with photos and technical drawings.

Understanding Boat Batteries and Battery Charging includes the following subjects: lead acid batteries;
AGM batteries; gel batteries; general battery information; battery ratings and selection; safety,
installation, and maintenance; charging, alternators, and regulators; and more.

UNDERSTANDING BOAT WIRING
John C. Payne

Understanding Boat Wiring includes the following subjects:
boat wiring standards; electrical school basics; system voltages; how to plan and install boat wiring;
circuit protection and isolation; switchboards and panels; grounding systems; and more.

"*...has a great deal of merit, offering concise, compact advice, with diagrams, on most electrical problems
that may occur on board. It is uncluttered with reams of technical jargon...*"
Sailing & Yachting

HOW TO INSTALL A NEW DIESEL ENGINE
Peter Cumberlidge

How to Install a New Diesel Engine covers all the factors relevant to choosing and installing a new
diesel engine, whether the work will be done by the owner or entrusted to a boatyard or engineer.

"*Highly recommended*"
Motorboat & Yachting

SHERIDAN HOUSE
America's Favorite Sailing Books
www.sheridanhouse.com

MARINE DIESEL ENGINES

Maintenance and Repair Manual

Jean-Luc Pallas

SHERIDAN HOUSE

This edition first published 2006
In the United States of America
by Sheridan House Inc.
145 Palisade Street
Dobbs Ferry, NY 10522
www.sheridanhouse.com

First published by Loisirs Nautiques 2004
under the title *Guide Pratique d'Entretien et
de Réparation des Moteurs Diesel*

Library of Congress Cataloging-in-Publication Data

Pallas, Jean-Luc.
[Guide pratique d'entretien et de réparation des moteurs
diesel. English]
Marine diesel engines maintenance and repair manual/
Jean-Luc Pallas.
p. cm.
ISBN 1-57409-236-7 (alk. paper)
1. Marine diesel motors—Maintenance and repair—
Handbooks, manuals, etc. I. Title.

VM770.P35 2006
623.87'2360288—dc22
    2006022429

SBN 10: 1-57409-236-7
ISBN 13: 978-1-57409-236-3

Designed by Paula McCann

Printed and bound in Spain by GraphyCems

Note: While all reasonable care has been taken
in the production of this publication, the publisher
takes no responsibility for the use of the methods
or products described in the book.

Contents

Repairs　143

Breakdowns　215

Winterising　221

Index　231

INTRODUCTION

TROUBLE-FREE CRUISING is every sailor's dream. So, to ensure that your holiday is not marred by mechanical glitches, make sure that your engine is well maintained. Many of the maintenance jobs are very quick and easy and, if done regularly, may save you trouble at sea.

How can this book help you?

It explains, in simple terms, how your boat's engine works and gives guidance on how to maintain and repair it.

Some of the jobs will need technical knowledge and ability, and special equipment, but the majority of tasks covered in the worksheets are within the ability of most boatowners who are interested in their engines and want to maintain engine performance without having to become an expert.

All these tasks, whether for maintenance or repair, are explained with precise illustrations which show the steps for each procedure. They are coded as either *simple*, *technical* or *complex*, depending on the level of skill and experience needed.

This book is divided into four parts. The first part covers engine theory in detail. In the second part, worksheets and checklists will help you to maintain your engine efficiently. The third part reviews the most common causes of engine breakdowns. A trouble-shooting list will help you to diagnose and fix them. Finally, the fourth part reviews the different steps to follow for one of the most important maintenance routines: winterising. Using the same step-by-step procedures as the worksheets, this section will show you how to lay up your engine to keep it in good condition throughout the winter.

THEORY

INVENTED AT THE END OF THE 19TH CENTURY, the diesel engine operates on the same principle as the internal combustion engine. Only the fuel and the intake phase differ. Before starting maintenance and repairs, it is sensible to learn a bit about your engine's anatomy – the different systems such as fuel, lubrication and cooling; or external systems such as the transmission, engine or electrical components. Explained in a simple and practical way, this section will help you to understand how your engine works.

Rudolf Diesel (1858–1913)

One September night in 1913, aboard the liner *Dresden* on the Calais to Dover run, a man fell overboard. His name on the passenger list is Rudolf Diesel.

Diesel, a name that has become part of everyday language, will forever be associated with the principles of diesel fuel-injected engines, for which he laid the foundation.

In 1887, Rudolf Diesel, born in Paris of German parents, began the study of the engine that bears his name. Ten years later, he built his first fuel-injected engine. At 5 tons and with 20 litres of displacement, this enormous vertical single cylinder engine produced 20hp at 170rpm. One peculiarity was its performance ratio: 26% – the best for any thermal engine. At the time, by comparison, the ratio for petrol engines was 20% and for steam engines, 10%.

Rudolf Diesel's theory

Based on the four-stroke petrol (gasoline) internal combustion engine's operating principles, the diesel engine is distinguished by the fact that when the intake valve opens on the intake stroke, the engine aspirates only air, unlike the petrol engine, which in its carburetted version aspirates air and fuel. When the air is compressed on the second stroke, the compression can reach 40 bars at 600°C. At the end of compression, diesel fuel is injected into the combustion chamber at high pressure. The high temperature in the combustion chamber causes the fuel to auto-ignite. The third and fourth strokes – combustion and exhaust – are identical in every respect to those of the four-stroke petrol engine.

1897: the first 20hp 'Diesel' engine.

The diesel principle

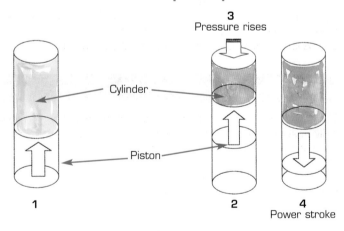

These 3 drawings show what happens in the engine cylinder

1 The piston traps a quantity of air.
2 The piston travels back up: the beginning of compression. The temperature of the highly compressed air rises.
3 End of compression; injection of diesel fuel under high pressure.
4 The increased pressure caused by the combustion of gases pushes the piston.

Diesel engine evolution

The diesel engines currently on the market operate by injecting pure diesel fuel. Earlier models used compressed air to inject fuel heated almost to its combustion point. The compression ratio was thus barely higher than in a petrol (gasoline) engine. As a result, these engines ran much more smoothly and quietly than those currently produced. The year 1910 marked an important date. The English engineer, Stuart MacKechnie, introduced his system of cold injection into highly compressed air. The very high compression ratio is what causes the characteristic knocking sound of today's diesel engine. But countless other improvements have been made: direct injection has given way to indirect injection into the 'pre-combustion chamber'. This solved some of the drawbacks associated with direct injection (knocking and lack of smoothness); the engine runs more smoothly with less noise. In 1990, for reasons of fuel economy and performance, direct injection made a comeback. Many improvements were made. Direct injection was refined and now gives, at the turn of the third millennium, peak performance for the diesel engine.

 The significant mechanical and thermal constraints found in this engine type require more robust components, capable of resisting higher pressures than those in a petrol (gasoline) engine. The moving parts (piston, connecting rod, crankshaft) are correspondingly oversized. Provided it is never subjected to demands greater than the manufacturer's design specifications, the diesel engine logically has a longer life than a petrol engine of similar power. Furthermore, the diesel engine's lack of an ignition system gives rise to fewer faults and has lower maintenance costs.

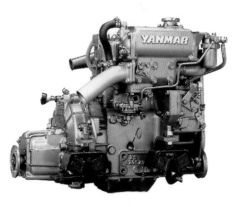

Robustness, longevity, better power and, lower pollution (resulting from more complete combustion) are the advantages of the diesel engine – making it the popular choice for engines used in our sailboats and other vessels with inboard engines.

Distribution networks

The majority of engines used in the boating industry were originally designed and mass-produced for land-based applications. Marinisation, which consists of adapting these engines to meet the demands of marine conditions, is done by the manufacturers. The modifications affect the cooling, exhaust, gearbox and electrical systems. However, some manufacturers have developed models exclusively for marine use: Volvo, Lombardini, Bukh.

Of the major manufacturers, four share the market for recreational boating: Volvo, Yanmar, Perkins, and Mercruiser for powerboats. They offer a choice of engines ranging from 8hp to more than 700hp. Nowadays, lesser known brands like Nanni, Vetus and Lombardini have gained a significant share of the market for replacement engines and for boats built partially or completely by amateurs.

THE PROPULSION SYSTEM

The inboard drive system comes in different forms. But the main distinction is between the stern tube shaft system and the S-Drive transmission system.

Even though the great majority of sailboats with inboard engines have stern tube shaft systems, manufacturers of 7 to 10 metre sailboats now tend to choose an S-Drive for their transmission system.

Depending on the type of transmission, the propulsion system is made up of three or four distinct parts:

◆ *Engine:* supplies the mechanical energy needed for propulsion.
◆ *Gearbox/reduction gear:* reduces the engine's revolutions and provides neutral, forward and reverse gears.
◆ *Stern tube shaft system:* comprises several components, ie the coupling, the shaft seal and the propeller shaft.
◆ *Propeller:* converts the engine torque into propulsive energy.

In an S-Drive transmission, the gearbox and propeller shaft are a single unit: the lower leg.

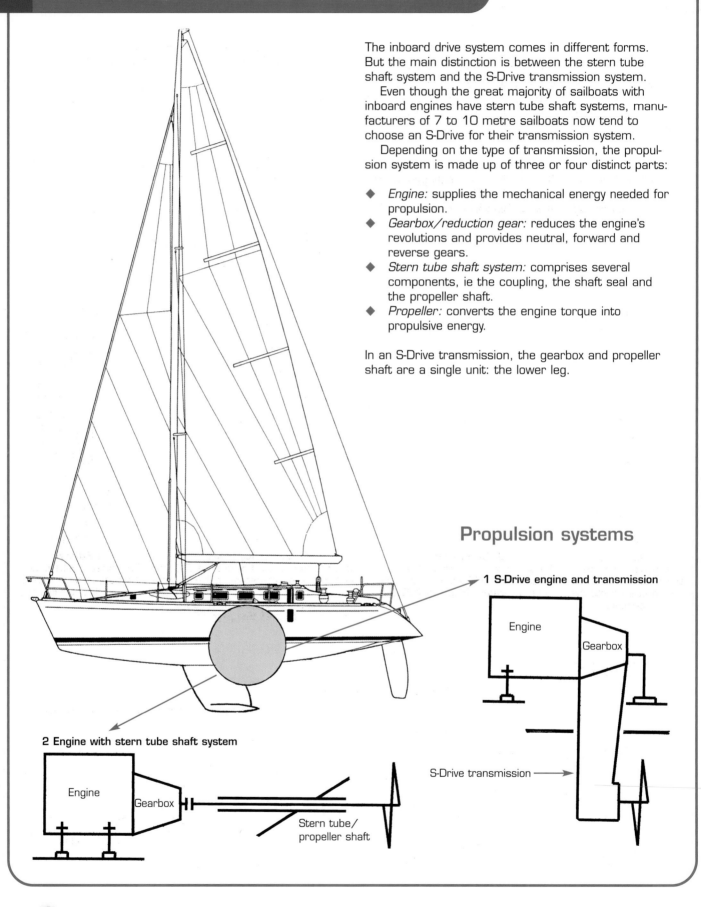

Propulsion systems

1 S-Drive engine and transmission

Engine

Gearbox

S-Drive transmission

2 Engine with stern tube shaft system

Engine

Gearbox

Stern tube/ propeller shaft

OPERATING PRINCIPLES

The diesel engine has four fundamental phases:

Induction – compression – ignition – exhaust

Depending on whether the cycle takes place in one or two crankshaft revolutions, the diesel engine is either a two-stroke (one crankshaft revolution) or a four-stroke (two crankshaft revolutions).

 Two-stroke diesel engines with specific power of up to 100hp per litre, are only produced for models over 200hp. Their production is currently limited to high-end power boats.

At the end of the compression stroke
Injection of diesel fuel under high pressure (170/250 bars).

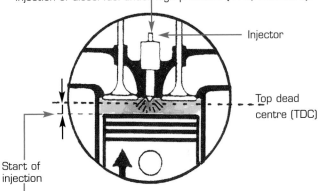

Injector

Top dead centre (TDC)

Start of injection

Four-stroke diesel engine cycle

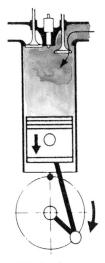

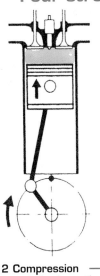

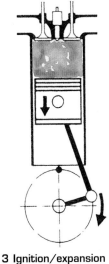

1 Induction
The intake valve opens; the piston travels down and draws air into the cylinder.

2 Compression
The piston travels back up; the highly compressed air (30 to 40 bars) rises in temperature (600° to 700°C).

3 Ignition/expansion
The diesel fuel injected before the TDC (top dead centre) spontaneously ignites on contact with the air. The expansion of the gas pushes the piston toward the BDC (bottom dead centre). See page 20.

4 Exhaust
The exhaust valve opens; the piston travels back up and expels the burnt gas.

Engine power cycle diagram

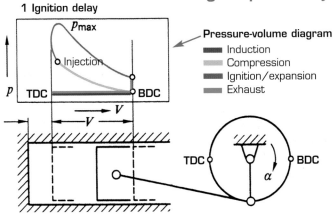

1 Ignition delay

p_{max}

Injection

p TDC BDC

V

V

TDC BDC

α

Pressure-volume diagram
- Induction
- Compression
- Ignition/expansion
- Exhaust

Pressure-crankshaft angle of rotation diagram

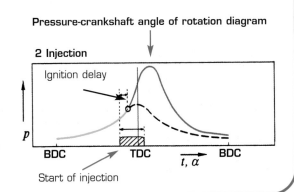

2 Injection

Ignition delay

p

BDC TDC BDC

t, α

Start of injection

Engine design

Engine power is itself directly related to the piston displacement and revs per minute (rpm). For a powerful engine, the manufacturer has two alternatives: increase the bore and the stroke, ie the piston displacement or, increase the rpm. But increasing the revs has its limits, due primarily to the mass of the moving parts. This is why manufacturers produce engines with multiple cylinders.

To increase power, it is therefore necessary to increase the number of cylinders, which makes it possible to regulate torque and reduce the mass of moving parts per cylinder.

Many manufacturers develop their power range starting with one cylinder as a reference point. Several sets of identical single cylinders then drive one single crankshaft. This is the case with Yanmar's GM series (subdivided into GM1, GM2, and GM3), or Volvo's 2000 series (2001, 2002, 2003) with 1, 2, or 3, corresponding to the number of cylinders.

3HM Engine – viewed from the induction/intake side

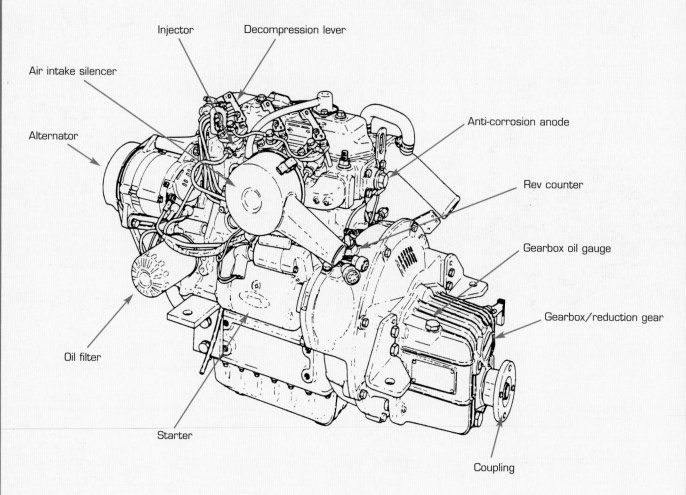

Injector

Decompression lever

Air intake silencer

Anti-corrosion anode

Alternator

Rev counter

Gearbox oil gauge

Gearbox/reduction gear

Oil filter

Starter

Coupling

In general, manufacturers use a single cylinder for engines under 10hp, two cylinders for 20hp, three cylinders for 30hp, and four cylinders for 40hp engines.

For higher power ranges, manufacturers increase the displacement of the reference cylinder, then go from four to sometimes five or six cylinders.

The cylinders have identical cycles but are offset in timing so that the strokes are spread evenly over the whole cycle.

While the connecting rod/piston assembly might be identical for a given series, this is not the case for the cylinder head, engine block or crankcase.

Cylinder arrangements

In-line cylinders

1 cylinder (under <10hp)

2 cylinders (± 20hp)

3 cylinders (± 30hp)

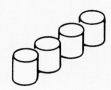

4 cylinders (± 90hp)

5 cylinders (± 300hp)

V cylinders

V6 and V8 cylinders: 350hp < 12000hp

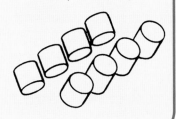

3HM engine – viewed from exhaust side

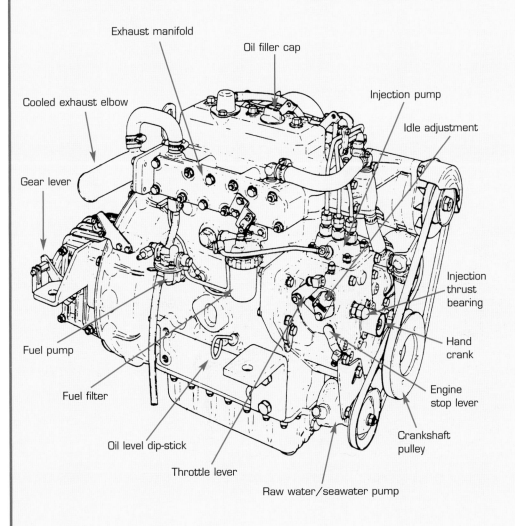

Exhaust manifold

Oil filler cap

Cooled exhaust elbow

Injection pump

Idle adjustment

Gear lever

Fuel pump

Fuel filter

Oil level dip-stick

Throttle lever

Raw water/seawater pump

Injection thrust bearing

Hand crank

Engine stop lever

Crankshaft pulley

Engine block – exploded view

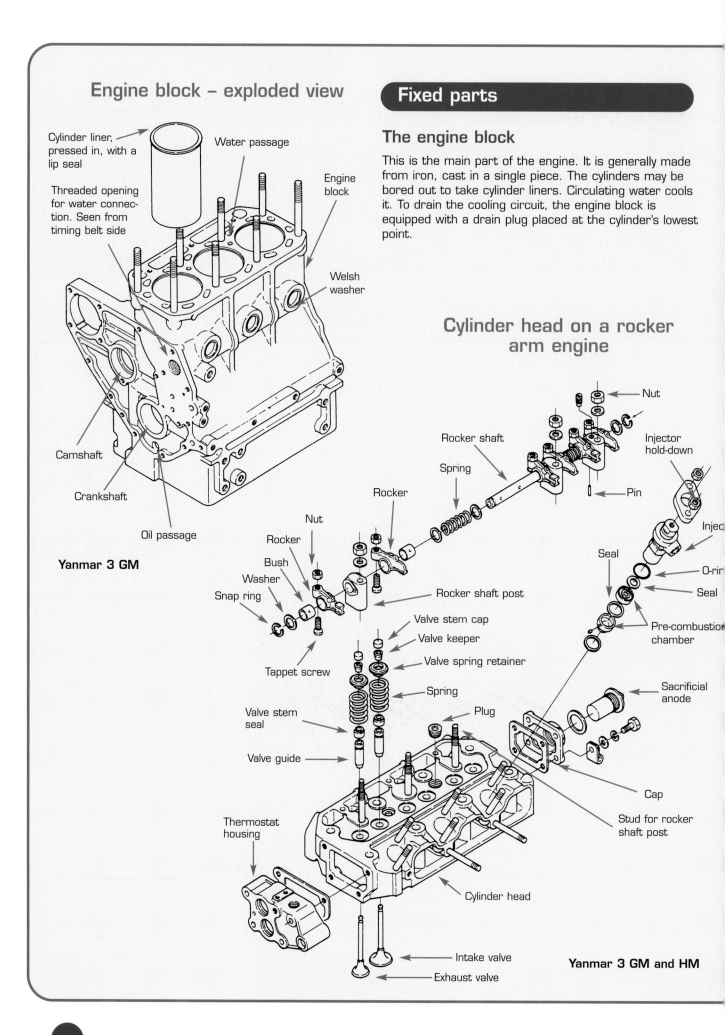

Cylinder liner, pressed in, with a lip seal

Water passage

Engine block

Threaded opening for water connection. Seen from timing belt side

Welsh washer

Camshaft

Crankshaft

Oil passage

Yanmar 3 GM

Nut

Rocker shaft

Spring

Rocker

Injector hold-down

Pin

Injec

Seal

Nut

Rocker

Bush

O-rir

Washer

Rocker shaft post

Seal

Snap ring

Pre-combustion chamber

Valve stem cap

Valve keeper

Tappet screw

Valve spring retainer

Spring

Sacrificial anode

Valve stem seal

Plug

Valve guide

Cap

Thermostat housing

Stud for rocker shaft post

Cylinder head

Intake valve

Yanmar 3 GM and HM

Exhaust valve

The engine block

This is the main part of the engine. It is generally made from iron, cast in a single piece. The cylinders may be bored out to take cylinder liners. Circulating water cools it. To drain the cooling circuit, the engine block is equipped with a drain plug placed at the cylinder's lowest point.

Cylinder head on a rocker arm engine

The cylinder head

Located at the upper end of the cylinder, it closes the cylinder and forms the combustion chamber. It contains the injectors together with the intake and exhaust ducts. As it is subjected to very high temperatures, water passages are essential for cooling.

The combustion chamber volume determines the compression ratio. The injection type (direct or indirect) and its means of distribution (two, three or four valves per cylinder) are factors in its design.

The head gasket

Generally composed of two copper foils with an insulation layer in between, it is sometimes reduced to its simplest form: a single copper sheet. The head gasket provides the seal between the cylinder head and the engine block.

The sump and covers

Oil sump, valve cover or engine front cover: made of pressed sheet metal or cast in light alloys, they constitute covers or plates that close different engine surfaces.

Cylinder head on an overhead cam engine

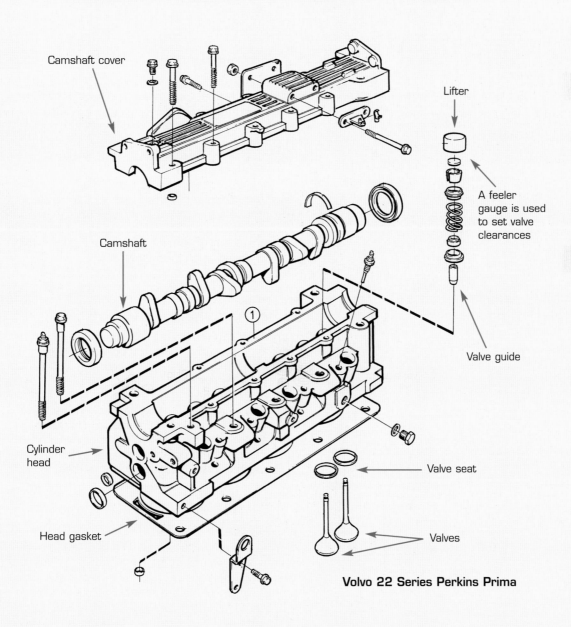

Camshaft cover

Lifter

A feeler gauge is used to set valve clearances

Camshaft

①

Valve guide

Cylinder head

Valve seat

Head gasket

Valves

Volvo 22 Series Perkins Prima

Engine structural design

The engine consists of fixed parts...

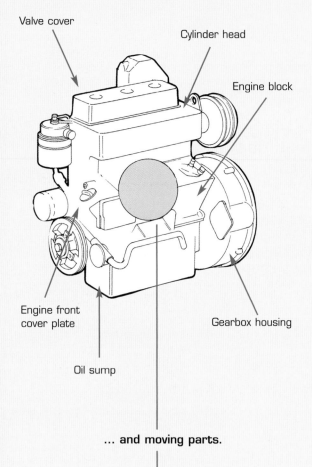

Valve cover

Cylinder head

Engine block

Engine front cover plate

Gearbox housing

Oil sump

... and moving parts.

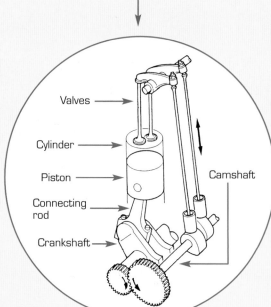

Valves

Cylinder

Piston

Connecting rod

Crankshaft

Camshaft

Moving parts

Torque is transmitted by means of a dynamic system containing three major elements: the piston, the connecting rod, and the crankshaft.

The piston

The piston is made of light alloy and has an alternating rectilinear motion. The piston crown forms part of the combustion chamber. For this reason, it is sometimes hollowed out with cavities designed to create turbulence to improve combustion.

Rings set in the upper part of the piston, towards the crown, make the combustion chamber airtight. They are known as the compression ring, the oil ring, and the scraper ring – which is usually located below the piston pin. The compression ring is often chromed. It is placed away from the piston edge to avoid direct exposure to the heat produced during combustion.

The connecting rod

The connecting rod makes the connection between the crankshaft and the piston. Cast in steel, it has to resist very high compression stress. For this reason, manufacturers have adopted an H-shaped section. The connecting rod's big end is often cut at an oblique angle to allow the connecting rod/piston assembly to be extracted through the top of the cylinder.

 The connecting rod's big end bearing cap is fitted to match the connecting rod's orientation. When reassembling the engine make sure to check the alignment of the assembly marks provided by the manufacturer.

The bearing shells

Comprised of two removable half shells coated with a layer of anti-friction metal, they make the contact between the crankshaft journals and the connecting rod.

 The wear marks seen on the bearing shells when dismantled are often caused either by a lack of oil or a lack of oil pressure. When performing a complete engine overhaul it is essential to check the entire oil circuit.

The crankshaft – flywheel assembly

Consisting of the crankshaft and the flywheel, it transfers the combustion energy in the form of torque. The flywheel balances the crankshaft rotation and makes the engine run smoothly. The precision-machined crankshaft is made from steel or nickel chrome – designed to withstand high temperature and carefully balanced – making this one of the engine's most important components.

Exploded view of connecting rod – piston assembly components

The piston grooves hold:

The compression ring

The contact surface
is chromed

The second compression ring

The ring is
tapered (note
which way)

The oil ring

The oil ring has an
expansion spring. With a
small friction surface, it
exerts high pressure on
the cylinder walls.

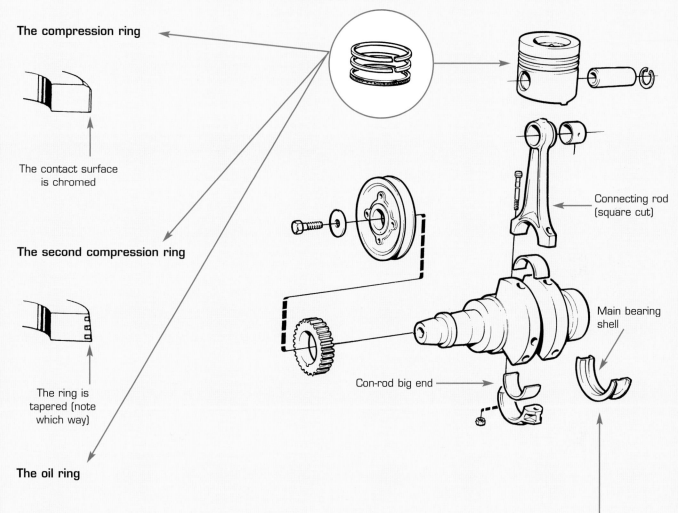

Connecting rod
(square cut)

Main bearing
shell

Con-rod big end

*The oblique angle on the
connecting rod fork allows the
piston/rod assembly to be
removed through the top of the
cylinder.

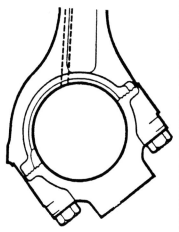

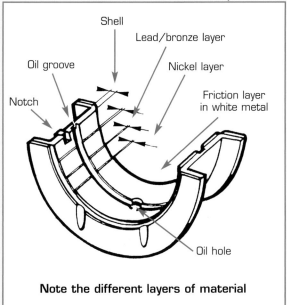

Shell

Lead/bronze layer

Nickel layer

Friction layer
in white metal

Oil groove

Notch

Oil hole

Note the different layers of material

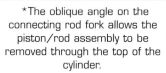

Different valve chain systems

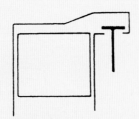

1 Side-valve engine
No longer in use. It was used in Renault Marine BD 1 and 2 engines.

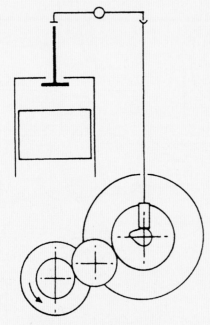

2 Overhead valve engine
Most frequently used system today. Yanmar, GM, Volvo 20, 23, 10, 30, Perkins 4108.

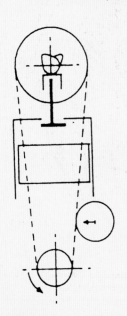

3 Overhead cam engine
Modern distribution solution allowing a reduction in the number of parts. The camshaft/ crankshaft connection is made by a timing belt. Used most notably in Perkins Prima engines or the Volvo 22 series.

Valve train

Air intake and evacuation of burnt gases are managed by valves. Their opening and closing is controlled by a mechanism which is very important for correct engine timing which we will call: the valve train.

The system is made up of a crankshaft, a means of connection, and in general, two valves per cylinder. The valves act like taps, opening and closing.

The camshaft

The camshaft is driven by the crankshaft and has as many cams as there are valves. Its location within the engine varies, depending on the design.

The most common arrangement on marine engines is the 'rocker arm' system. The camshaft is located in the engine block and is driven by a set of gears with a 50% reduction ratio. A set of lifters, pushrods, and rocker arms provides the connection between the camshaft and valves. Coil springs around the valves close them automatically when the pressure from the cam ceases. When the camshaft is in the cylinder head, it is called an 'overhead cam' engine. This type reduces the number of components, thereby reducing the engine weight. There are no lifters, push rods and rocker arms; a timing belt provides the camshaft/ crankshaft connection. This modern concept has several advantages: reduction of mass in motion, elimination of the connection system and its need for lubrication, plus quiet operation.

The valves

Depending on the engine's design and horsepower, there are generally two valves per cylinder: one for intake; one for exhaust. To improve induction and exhaust, some engines may have three or even four valves per cylinder. Each valve is composed of a head with a bevelled edge and a stem to guide it.

Subjected to very rapid alternating movement, the valve heads deteriorate, the air-tightness of the combustion chamber is compromised and starting problems and loss of power begin to appear. It is time to regrind valves and seats.

The rocker arms

Sometimes called tappets, the rockers transmit the movement of the cams to the valves by way of pushrods. The end of the rocker arm has a nut/screw system to adjust the rocker clearance gap.

Overhead cam system

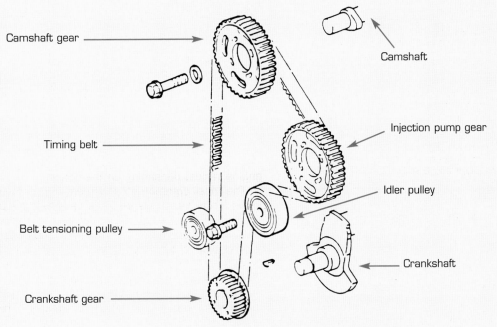

Camshaft gear

Camshaft

Injection pump gear

Timing belt

Idler pulley

Belt tensioning pulley

Crankshaft

Crankshaft gear

The timing belt should be changed every 2000 hours or every 30 months. (See engine manual.)

Rocker arm system

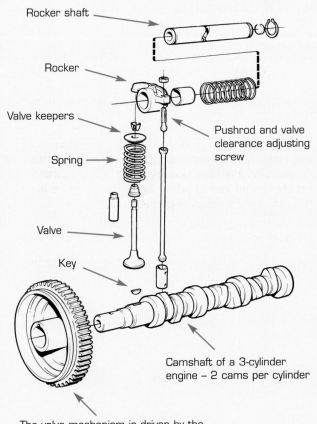

Rocker shaft

Rocker

Valve keepers

Pushrod and valve clearance adjusting screw

Spring

Valve

Key

Camshaft of a 3-cylinder engine – 2 cams per cylinder

The valve mechanism is driven by the main gear on the crankshaft

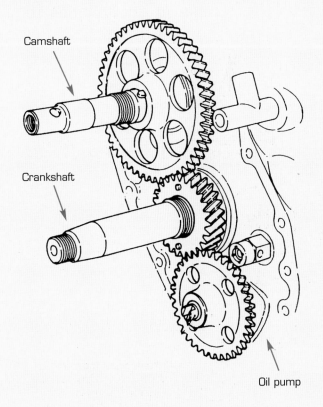

Camshaft

Crankshaft

Oil pump

Camshaft in the engine block
Note the valve chain reduction ratio: the camshaft turns at half the speed of the crankshaft.

Engine characteristics

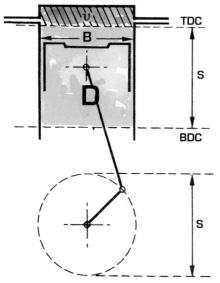

B Bore
S Stroke
D Displacement
CR Combustion
chamber volume

$$D = \frac{B^2 \times \pi \times S}{4}$$

$$\vartheta = \frac{B}{CR - 1}$$

Commonly used terms

The manufacturer's manual is filled with data and technical terms. A certain level of knowledge is needed to use it effectively. Here are some simple definitions:

Bore: diameter of the cylinder.

Top dead centre (TDC): the piston's uppermost position or the end of the upward stroke.

Bottom dead centre (BDC): the piston's lowest position or the end of the downward stroke.

Stroke (S): the distance travelled by the piston between top dead centre and bottom dead centre. It corresponds to a half turn of the crankshaft, ie: 180°.

Displacement volume: the difference between the volume swept by the piston and total volume.

Volume swept by the piston: the volume displaced by the piston between the bottom dead centre and the top dead centre in cubic centimetres.

Total volume: the volume swept by the piston multiplied by the number of cylinders. It is an essential characteristic of the engine.

Table of equivalents for old and new units of measure

	New unit SI System	Multiples (no longer in use)	Old units	Equivalent	Observations
Force	**Newton** (N)	dekanewton (daN)	kilogram-force (kgf)	1kg ~ 9.8N 1kgf ~ 0.93daN	1daN 10N Incorrect:1kg
Torque (for tightening)	**Newton-metre** (Nm)		[kilogramme force metre (kgf.m)]	1kgf.m ~ 9.8Nm 1kgf.m ~ 0.98daNm	eg 10kgf.m ~ 98Nm or 10daNM at 2% more or less
Energy (eg quantity of heat)	**Joule** (J)	kilojoule (KJ)	kilogram-force metre (kgf.m)	1kgf.m ~ 9.8J	1 calorie = 4.1855J 1 watt-hour = 3600J
Power	**Watt** (W)	kilowatt (kW)	horsepower (hp)	1hp ~ 736W 1hp ~ 0.736kW	1 horse (hp): old unit 'horsepower'
Pressure or stress (material resistance)	**Pascal** (Pa) or **Newton** per square metre (N/m^2)	bar (bar) hectobar (hbar)	kilogram-force per square centimetre (kgf/cm^2) kilogram-force per square millimetre (kgf/mm^2)	1 bar = 100 000Pa 1 hbar = 100 bars 1 kgf/cm^2 ~ 0.98 bar 1kgf/mm^2 ~ 0.98 bar	eg 10kgf/cm2 ~ 9.8 bars Atmospheric pressure: 101.325Pa 1.013 millibars 1.013 bar
Temperature	Degrees **Kelvin** (°K) **Celsius** (°C)				The old degrees Celsius denominations were: Degrees centigrade, then centesimals of degrees 0°C = 273.15°K
Mass	**Kilogram** (kg)	ton (t) gram (g)	kilogram-weights (kgw)	1kgw = 1kg	We no longer refer to body 'weights', but to their mass

Note: the mathematical symbol ~ means 'approximately equal to'

Power: traditionally expressed as horsepower. This now tends to be superseded by the European measure kilowatt. It indicates the power the engine can put out at a given number of rpm. Starting from torque values measured on the bench, the manufacturer calculates the power output per rpm for each engine type. The power ratings given by manufacturers are based on measurements made at the gearbox in accordance with ISO 8665 norms.

Compression ratio: the ratio between the total volume of the cylinder when the piston is at bottom dead centre and the volume remaining when the piston is at top dead centre.

Torque: the product of the force on the connecting rod times the length of the crank throw. This torque is measured in Newton-metres. It describes the maximum force produced by the engine at a given number of rpm. The greater the maximum torque at low rpm, the more smoothly the engine will run and vice-versa.

Specific fuel consumption: the mass of fuel consumed during a unit of time, or, the quantity of fuel in grams needed by the engine to produce 1W/h. The efficiency of inboard engines with the latest technology approaches 50%. Specific fuel consumption ranges between 160 and 210gr/hp/h.

Specifications
(Volvo MD 22 engine)

Model number MD22P
Power at the flywheel [2, 3, 4] kW (hp) 43.6 (59)
Power at the propeller shaft [2 3 4]
kW (hp) 41.9 (57)
rpm 3600–4000
Displacement (litres) 2.0
Number of cylinders .4
Bore (Nm) 84
Stroke (Nm) 89
Compression ratio 18:1
Engine weight dry with gearbox MS2A/MS2L . . .238
Weight, dry with transmission 120S (kg) 246

[1] Power at the flywheel in compliance with ISO 8665 or SAE J1228 standards
[2] Power at the propeller shaft in compliance with ISO 8665 standard or standards compatible with SAE J1228 and ICOMIA 28-83
[3] Nominal power in compliance with NMMA procedure
[4] With MS2

Characteristic engine curves
Volvo MD 22

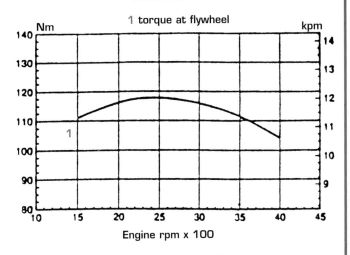

1 torque at flywheel

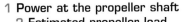

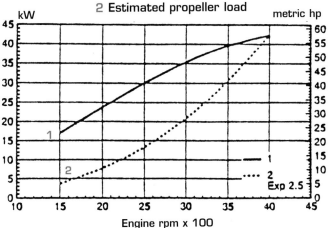

1 **Power at the propeller shaft**
2 **Estimated propeller load**

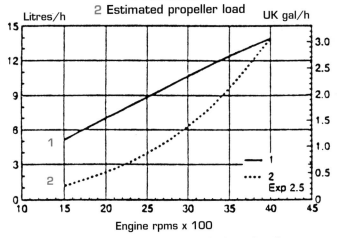

1 **Full throttle**
2 **Estimated propeller load**

A review of the manufacturer's published graphs allows us to:
- know the engine's power, torque and specific fuel consumption relative to rpms at full throttle;
- analyse the engine's performance at different rpm;
- identify the ideal rpm;
- compare different engines by analysing the curves.

Diesel engine injection: chamber types

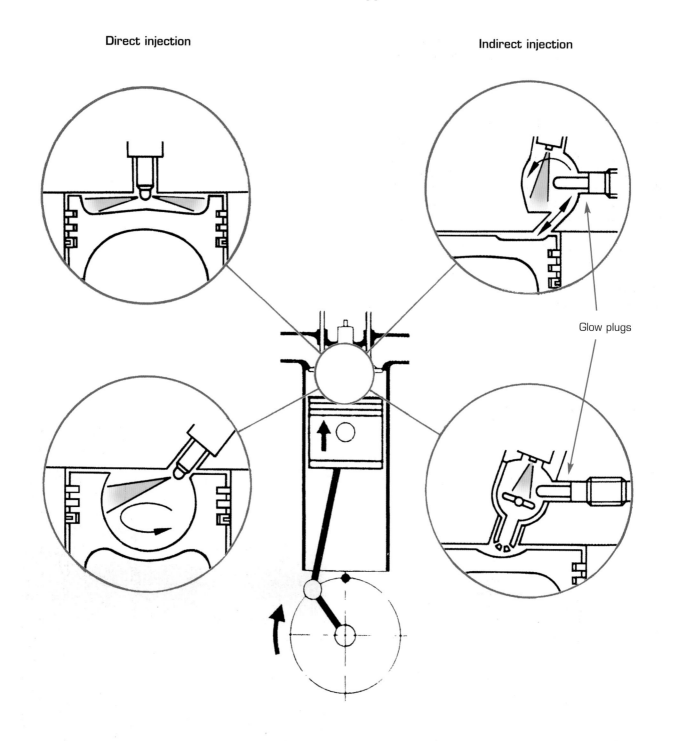

Direct injection

Indirect injection

Glow plugs

Injection begins just before TDC

DIESEL ENGINE TYPES

Diesel engines are classified into two categories based on how the fuel is injected: *direct injection engines* and *indirect injection engines*.

Direct injection engines

On this type of engine, a fuel injector nozzle with several jets protrudes directly into the combustion chamber. The piston might be flat or have a cavity depending on whether or not a turbulence effect for the fuel mix is desired. The turbulence of the compressed air together with the injected diesel fuel achieves fuel combustion.

    The compression ratio is very high and so is the injection pressure. The instant combustion of the diesel/air mix produces a maximum high pressure. This results in rough running. On the other hand, the specific fuel consumption is low and this type of engine does not require an auxiliary system to assist starting.

 Particular attention must be paid to the injection timing adjustment. Too advanced and the engine will knock excessively, risking damage to the moving parts. Too retarded and the engine will lose power.

Direct injection

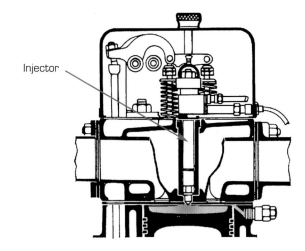

Injector

Note the injector location. It enters the chamber directly above the piston.

The pre-combustion/ swirl chamber

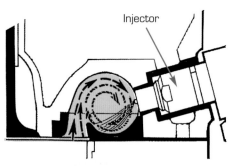

Injector

The solid arrows indicate pure air, the dotted ones show the fuel/air mix.

In order to remedy the defects caused by direct injection, ie roughness, knocking, lack of smoothness, etc, manufacturers developed an indirect injection engine. Here the injector delivers its charge into a pre-combustion or swirl chamber, the volume of which is part of the combustion chamber. This arrangement allows the use of a lower compression ratio as well as lower injection pressure. This engine runs much more smoothly than a direct injection engine and knocking is reduced. The only disadvantage for an indirect engine is slightly higher fuel consumption than that of a direct injection engine and the need for glow plugs to start the engine because the compression ratio is insufficient for auto-ignition when the engine is cold.

Pre-heating system

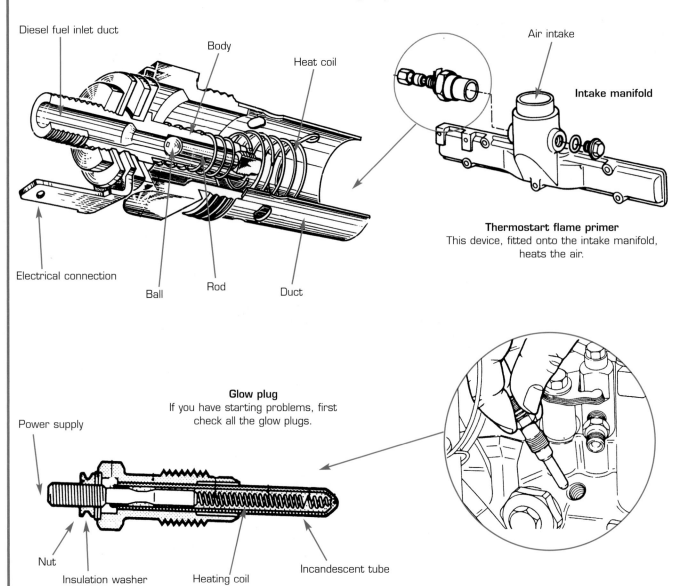

Diesel fuel inlet duct

Body

Heat coil

Air intake

Intake manifold

Electrical connection

Ball

Rod

Duct

Thermostart flame primer
This device, fitted onto the intake manifold, heats the air.

Power supply

Glow plug
If you have starting problems, first check all the glow plugs.

Nut

Insulation washer

Heating coil

Incandescent tube

FUEL/AIR SUPPLY IN DIESEL ENGINES

The fuel/air supply in diesel engines consists of two circuits: the air supply circuit and the diesel fuel supply circuit.

Air supply

Naturally aspirated engines

An engine is said to be naturally aspirated when the engine itself draws in air by means of the vacuum created when the piston travels down during the intake stroke. This method of aspiration is most commonly found in small marine diesel engines.

Supercharged or turbocharged engines

To force more air into the cylinder during the intake phase, some engines are equipped with a supercharger or turbocharger. These devices increase the specific mass of air by pre-compressing it.

Natural aspiration

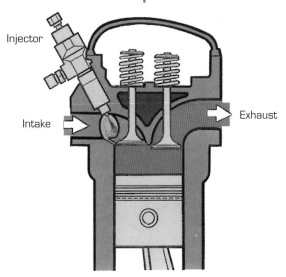

Injector

Intake

Exhaust

Super/turbocharging operating principles

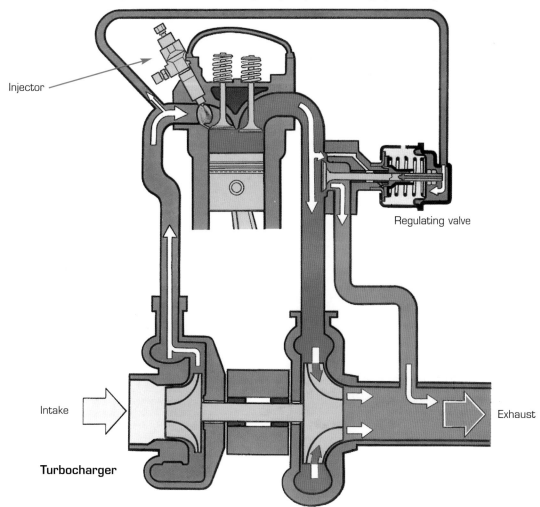

Injector

Regulating valve

Intake

Exhaust

Turbocharger

How the turbocharger works

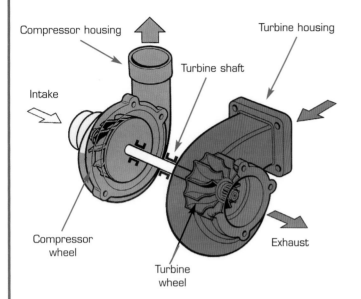

Compressor housing

Turbine housing

Turbine shaft

Intake

Compressor wheel

Exhaust

Turbine wheel

Super/turbocharging provides:

◆ increased engine power for the same displacement;
◆ improved engine performance at high revs and under heavy load.

The turbocharger uses exhaust gas energy. This energy is transferred by a pair of turbines. The drive turbine, which is driven by the exhaust gases as they exit the engine, drives the boost turbine. This turbine draws in outside air and pushes it upwards to the intake valve.

The super/turbocharge pressure is controlled by a regulating valve. This device, which rotates at very high speed (up to 200,000rpm), needs lubrication by pressurised oil. The high supercharging pressure requires a reduction in the compression ratio.

To improve engine performance across a range of uses, some marine engines are equipped with a supercharger, which performs well at high revolutions, connected to a compressor to enhance airflow at low revolutions. The compressor is driven mechanically by a belt on the crankshaft. This mechanical compressor gives immediate response to acceleration and provides significant torque at low revolutions. Engagement and disengagement are controlled by a computer that continuously analyses the engine parameters, particularly the load.

The turbocharger unit

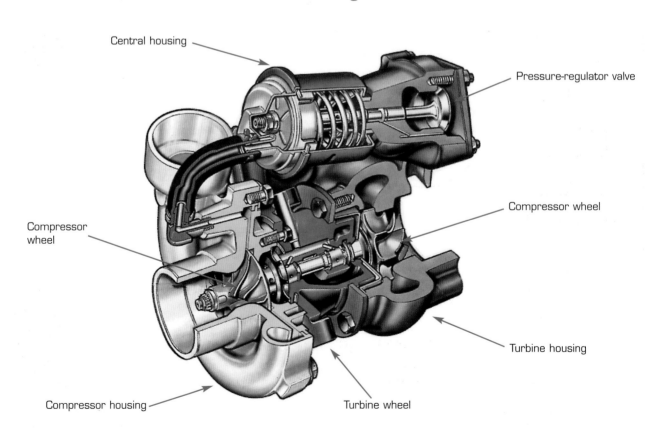

Central housing

Pressure-regulator valve

Compressor wheel

Compressor wheel

Turbine housing

Compressor housing

Turbine wheel

Engine fuel supply

The role of the fuel system is to feed each cylinder a set amount of clean filtered fuel, precisely measured, under high pressure at regular intervals – irrespective of engine operating conditions. Obviously, this is no easy matter.

Fuel system organisation

The fuel system of your boat's diesel engine includes:

- a tank,
- a primary filter/water separator,
- a filter,
- a fuel pump,
- one or more injection pumps,
- one or more injectors.

These components are connected by specific pipelines.

The diesel fuel circuit

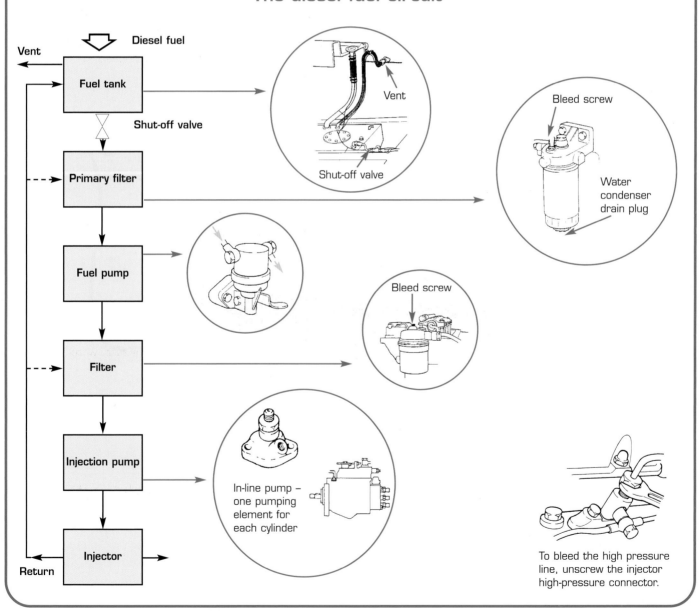

Vent

Diesel fuel

Fuel tank

Shut-off valve

Primary filter

Fuel pump

Filter

Injection pump

Injector

Return

Vent

Shut-off valve

Bleed screw

Water condenser drain plug

Bleed screw

In-line pump – one pumping element for each cylinder

To bleed the high pressure line, unscrew the injector high-pressure connector.

The fuel-tank

To avoid corrosion, the fuel tanks fitted on pleasure boats are built of stainless steel. Regulations require baffles in tanks over 45cm long to prevent the fuel from sloshing around.

The water-separating primary filter

This filter is mounted in-line between the tank and the fuel pump. Its role is to trap impurities ($1/100$ of a millimetre) and separate the water in suspension in the fuel by decantation. In reality, a boat's fuel tank contains a lot of water, which comes in with the fuel itself and is also produced by condensation.

The filter

Mounted in-line between the fuel pump and the injection pump, its role is to trap the smallest impurities (two to three microns) to protect the injection pump.

The filter or cartridge must be replaced periodically – approximately every 200 hours or every year.

The fuel system must be bled after any work has been carried out on it.

Diesel fuel circuit (ISO 7840 standard)

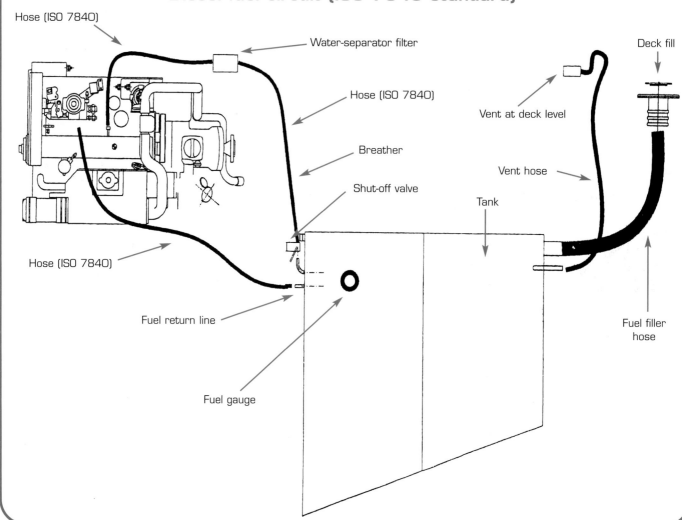

Hose (ISO 7840)

Water-separator filter

Deck fill

Hose (ISO 7840)

Vent at deck level

Breather

Vent hose

Shut-off valve

Tank

Hose (ISO 7840)

Fuel return line

Fuel filler hose

Fuel gauge

The fuel pump

Situated on the engine, it feeds fuel under light pressure to the injection pump.

The fuel pump is made up of a diaphragm and two valves: one inlet and one outlet. It is operated by the engine camshaft.

Diaphragm fuel pump

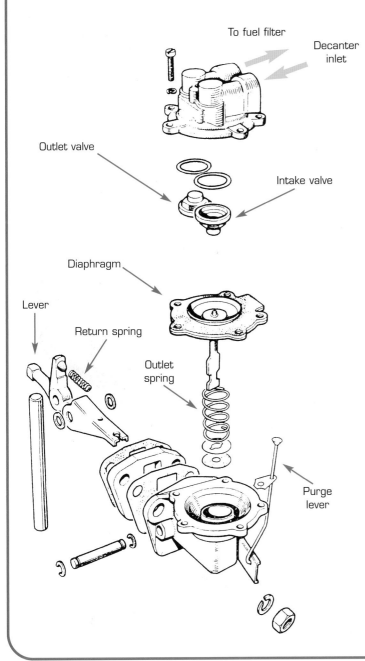

To fuel filter

Decanter inlet

Outlet valve

Intake valve

Diaphragm

Lever

Return spring

Outlet spring

Purge lever

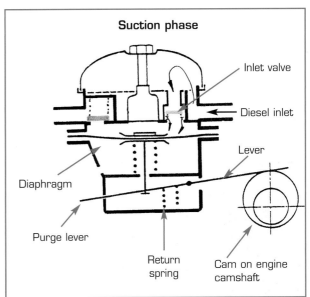

Suction phase

Inlet valve

Diesel inlet

Lever

Diaphragm

Purge lever

Return spring

Cam on engine camshaft

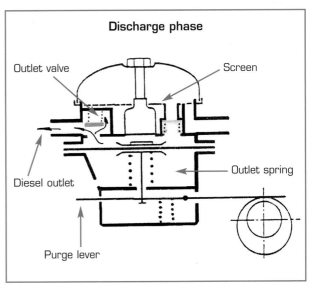

Discharge phase

Outlet valve

Screen

Diesel outlet

Outlet spring

Purge lever

The injection pump

The major components of the diesel engine are the injection pump and its injector which has the job of spraying a quantity of diesel fuel, corresponding to the power required by the user, into each cylinder at the end of the intake stroke.

The choice of injection pump type depends mainly on the number of cylinders.

Engines with 1, 2 or 3 cylinders have an injection or jerk pump equipped with as many pumping elements as there are cylinders. For four cylinder engines and above, manufacturers use an injection pump with a single pumping element. This type of pump is often called a rotary or distributor pump.

Diesel fuel circuit with rotary/distributor pump assembly

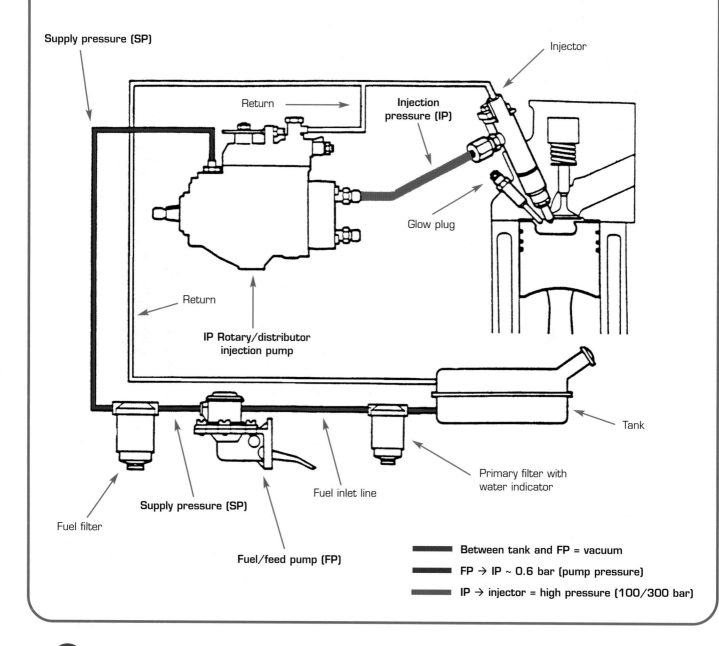

Supply pressure (SP)

Injector

Return

Injection pressure (IP)

Glow plug

Return

IP Rotary/distributor injection pump

Tank

Primary filter with water indicator

Fuel inlet line

Supply pressure (SP)

Fuel filter

Fuel/feed pump (FP)

Between tank and FP = vacuum

FP → IP ~ 0.6 bar (pump pressure)

IP → injector = high pressure (100/300 bar)

How the injection pump works

Remember that the pumping unit has two functions: to vary the rate of flow and to pressurise it.

In the case of injection pumps with one pumping element for each cylinder, the system, which is driven by a cam, consists of a plunger sliding in a barrel. During its stroke, the plunger successively opens the diesel fuel inlet and outlet ports. The plunger stroke is constant. The variation of the flow and thus the engine's power, is obtained by rotating the plunger to alter its travel.

Exploded view of an injection pump showing the pumping element and plunger position for different injection flow rates

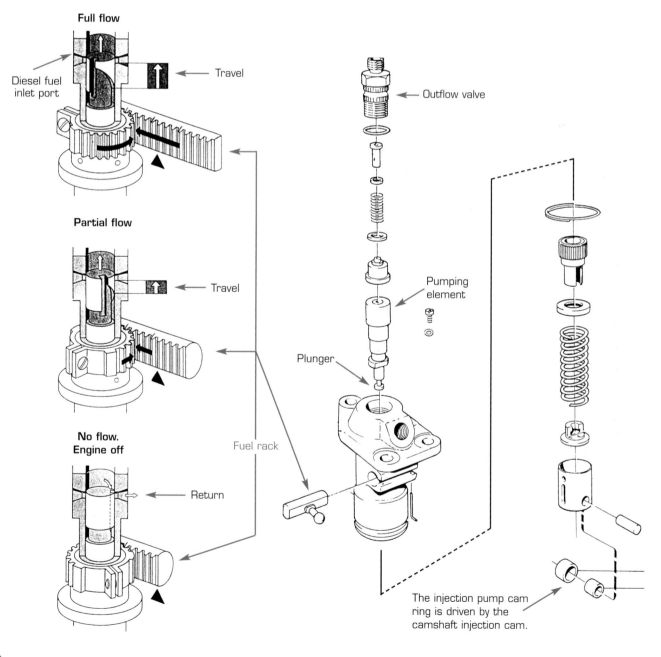

Full flow

Diesel fuel inlet port

Travel

Partial flow

Travel

No flow. Engine off

Return

Fuel rack

Outflow valve

Pumping element

Plunger

The injection pump cam ring is driven by the camshaft injection cam.

Distributor/rotary injection pump

The compact distributor or rotary injection pump was developed for smaller, faster, engines. Unlike the injection pump previously described, it has only one single pumping element for all of the cylinders and a rotary distributor that distributes fuel to each cylinder.

A rotating ring with as many cams as there are cylinders creates pressurisation by alternately displacing two opposing pistons. The fuel is then distributed by the rotary distributor. A transfer pump mounted on the far end of the distribution rotor feeds the pumping chamber. The position of a slide regulator determines the piston travel and thus the fuel flow rate. Stop screws determine the travel of the acceleration lever from maximum acceleration to idle.

In addition to performing the two functions described above, each injection system has other devices to allow:

◆ variation of injection timing according to engine acceleration;
◆ regulation of engine acceleration;
◆ stopping the engine;
◆ cold starting.

Injection pump: distributor/rotary type

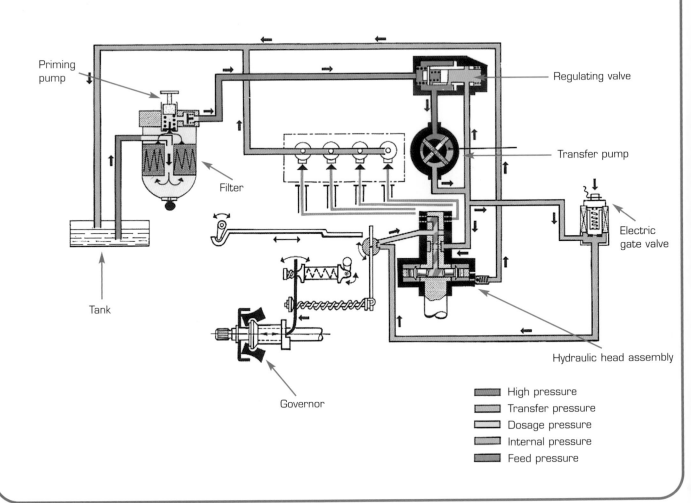

Cross-section of a distributor/rotary pump

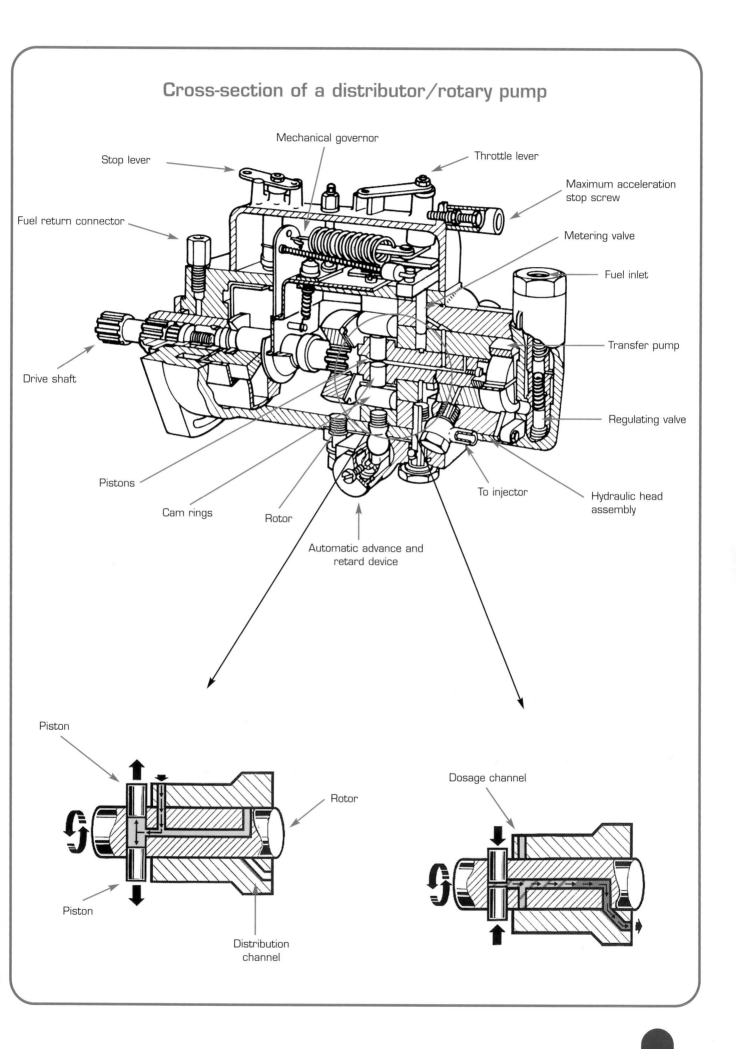

Stop lever

Mechanical governor

Throttle lever

Maximum acceleration stop screw

Fuel return connector

Metering valve

Fuel inlet

Transfer pump

Drive shaft

Regulating valve

Pistons

Cam rings

Rotor

Automatic advance and retard device

To injector

Hydraulic head assembly

Piston

Rotor

Piston

Distribution channel

Dosage channel

Fuel injection timing

In most cases, fuel injection timing is provided by a mechanical system. The injection must first be timed statically to a value between 3° and 5°. It can reach 25° or 30° (on average) at maximum acceleration.
Each device comprises:

◆ a mechanism that detects engine rpm;
◆ a mechanism that controls when injection starts.

The devices vary with the type of pump. They use either centrifugal force on a set of flyweights, or, in the case of the distribution/rotary pumps, the action of fuel pressure on a hydraulic ram.

Governors

Every injection system has a device to limit engine rpm by varying the amount of fuel injected in the engine. It is usually known as a governor. This device, which responds to load and rpm, is usually a centrifugal type. Generally incorporated in the injection pump, it is often separate from it on 1, 2 and 3 cylinder engines.

Adjusting injection timing

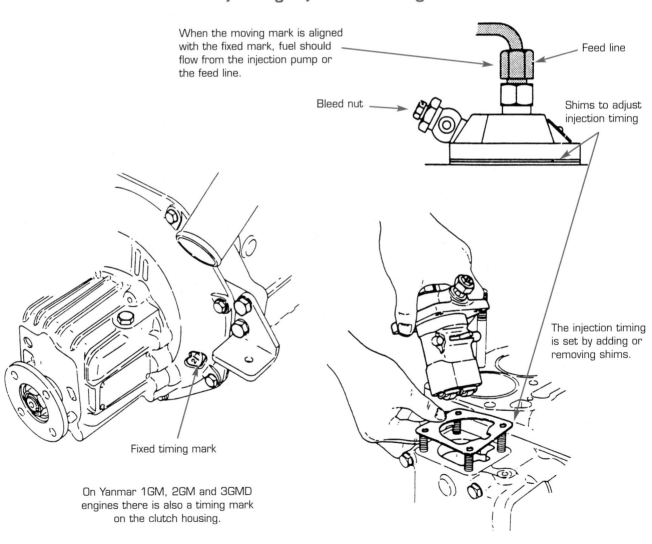

When the moving mark is aligned with the fixed mark, fuel should flow from the injection pump or the feed line.

Feed line

Bleed nut

Shims to adjust injection timing

The injection timing is set by adding or removing shims.

Fixed timing mark

On Yanmar 1GM, 2GM and 3GMD engines there is also a timing mark on the clutch housing.

Engine shut-off

The engine is shut off by a mechanical or electrical device that cuts off the fuel supply.

Cold starting

Cold starting a diesel engine is difficult because of how it works. Heat loss and leaks reduce the pressure and temperature at the end of compression so much that starting is impossible without auxiliary devices.

There are different ways of cold starting, depending on whether injection is direct or indirect.

In the case of direct injection, no device may be needed or it may be reduced to a simple injection overfeed.

Some older engines are equipped with an air heater fitted on the inlet manifold.

Engines with indirect injection need the help of glow plugs to start. These can be controlled automatically by electronic modules that take the ambient air and engine temperatures into consideration.

Injection pump governor

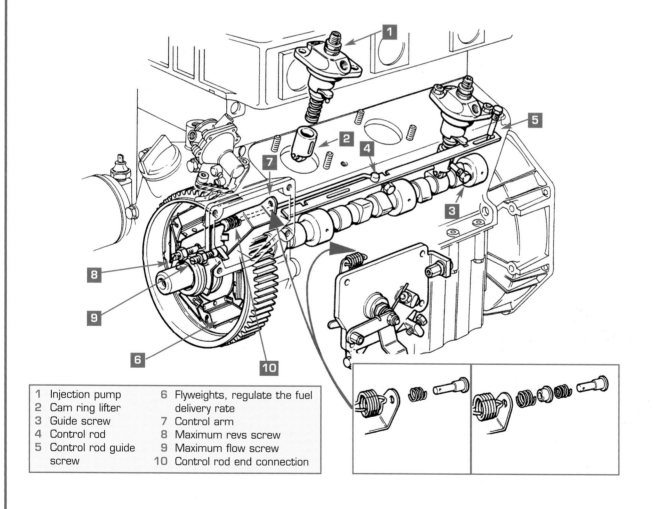

1	Injection pump	6	Flyweights, regulate the fuel delivery rate
2	Cam ring lifter	7	Control arm
3	Guide screw	8	Maximum revs screw
4	Control rod	9	Maximum flow screw
5	Control rod guide screw	10	Control rod end connection

Injector

Located on the cylinder head, it either enters directly into the combustion chamber (direct injection) or into a pre-combustion chamber (indirect injection). Its role is to atomise the diesel fuel fed under high pressure by the injection pump and to spray it in the combustion chamber. The injector is fitted in the nozzle holder with a threaded fitting and comprises a body and a needle. The shape of the needle end depends on the type of spray pattern desired.

The needle valve is kept in its seat by a spring which is set to a precise degree of pressure. The injection pressure is regulated by a set of shims or an adjusting screw.

Injector types and spray patterns

Each injector has its own characteristics of injection pressure and spray pattern shape and direction, depending on the type of engine it is fitted in. We mainly distinguish between the hole type of injector nozzle, designed for direct injection engines, and the pintle injector for indirect injection engines.

Pintle injector

The injector has a single fuel outlet. The end of the needle is visible depending on its shape; the spray will be more or less directional.

Hole type nozzle

The fuel is injected in several jets to achieve perfect atomisation. The number of holes and their sizes vary depending on the shape of the combustion chamber.

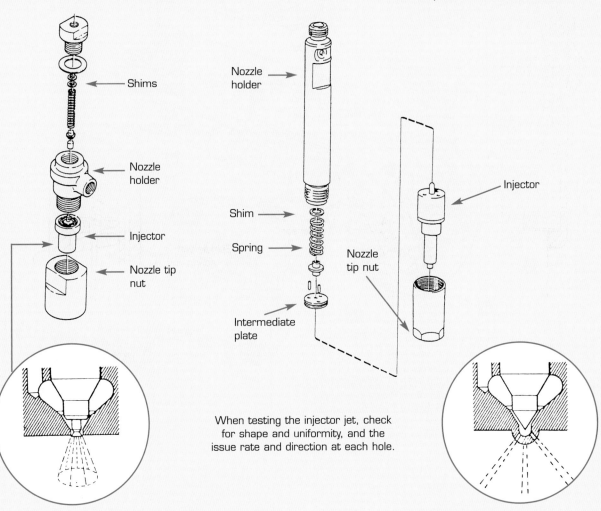

When testing the injector jet, check for shape and uniformity, and the issue rate and direction at each hole.

Principles of operation

The diesel fuel coming from the injection pump enters at the base of the needle and flows into the pressure chamber. When the fuel pressure exceeds that exerted by the spring, the needle lifts and fuel is then sprayed at high pressure into the cylinders. When the injection pump stops feeding fuel, the pressure becomes lower than that of the spring and the needle falls back to its seat. This is the end of the fuel injection process.

The fuel, which seeps through the length of the needle valve stem, lubricates the needle then returns to the fuel tank through a leak-off pipe.

Electronic sensors on nozzle holders with an injection start indicator

One of the current developments of injection pump electronics requires sender units that transmit information to the control box. One of these sender units is located in each nozzle holder and transmits information to the electronic control box:

◆ Injection start
◆ Engine rpm
◆ Injection duration

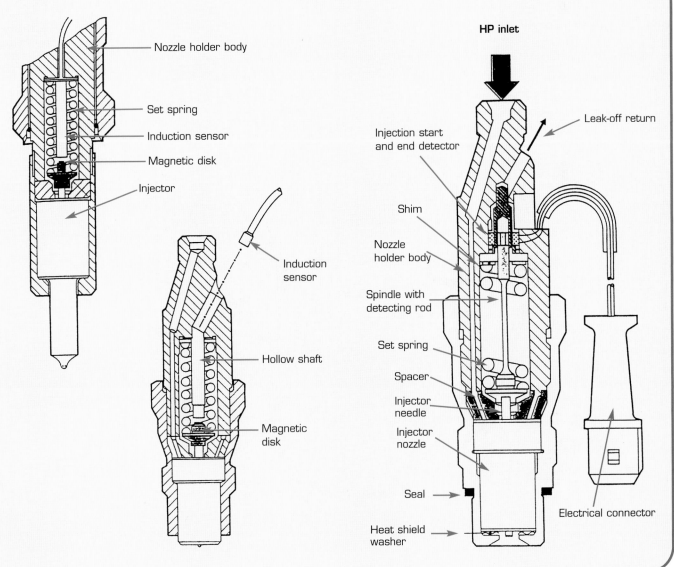

LUBRICATION

The engine lubrication system is designed to reduce friction between the moving parts by coating them with a thin film of oil. It also flushes away particles of metal and carbon deposits.

Lubrication also helps to cool the engine components and contributes to the air-tightness of the combustion chamber.

Absence of lubrication causes a rise in friction temperature that, over time, will cause the piston/ connecting rod/crankshaft assembly to seize.

Lubrication system

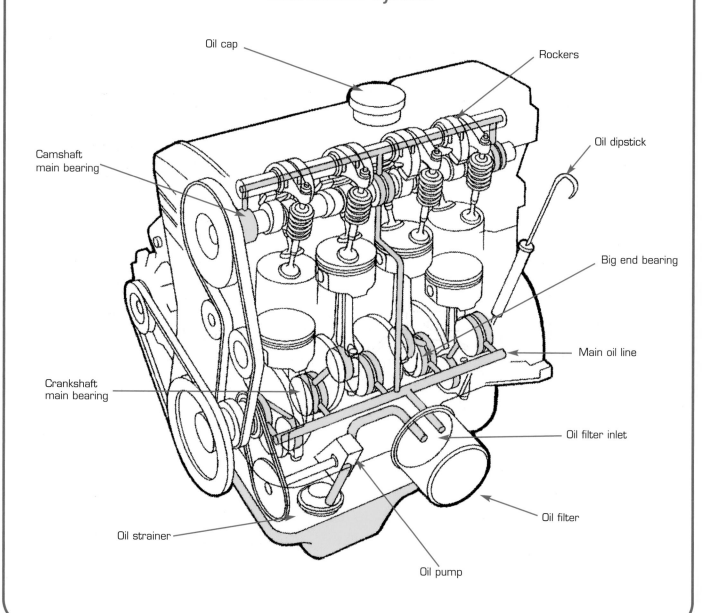

Oil cap

Rockers

Oil dipstick

Camshaft main bearing

Big end bearing

Main oil line

Crankshaft main bearing

Oil filter inlet

Oil filter

Oil strainer

Oil pump

Problems with the lubrication system rarely arise; use good quality oil, check your oil level often and carry out regular services as recommended by the manufacturer.

The lubrication circuit

While some older engines still use splash lubrication, almost all modern marine engines are lubricated by oil circulating under pressure.

Engine oil circuit

Perkins Prima – Volvo 22 series

1 Oil pump
2 Pressure release valve
3 Oil cooling
4 Oil filter
5 Crankshaft
6 Camshaft

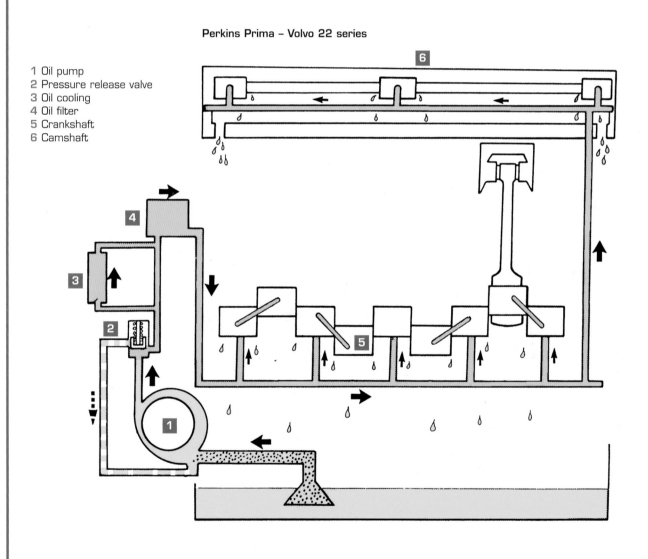

Pressurised lubrication

The system consists of:

◆ An oil pump that draws in oil from the sump and sends it under pressure into the line.
◆ A release valve that regulates the oil pressure irrespective of engine rpm.
◆ A filter to purify the oil.
◆ A by-pass integrated in the filter, which allows oil to circulate when the oil filter is clogged.
◆ A set of lines that distribute the oil under pressure to the components to be lubricated. The oil then falls back into the oil sump by gravity through the return ducts.
◆ An oil pressure sensing device, called an oil pressure sender unit, is mounted in by-pass position on the oil line. This tells you the oil pressure in the circuit by means of a warning light or a pressure gauge. The system is also connected to an audible alarm.

Pressurised lubrication system – Yanmar 3GM and 3HM engines

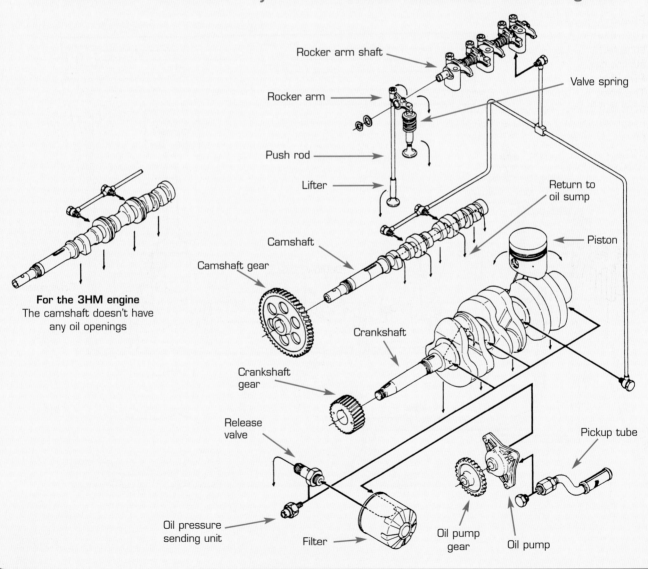

For the 3HM engine
The camshaft doesn't have any oil openings

Rocker arm shaft
Valve spring
Rocker arm
Push rod
Lifter
Return to oil sump
Camshaft
Piston
Camshaft gear
Crankshaft
Crankshaft gear
Release valve
Pickup tube
Oil pressure sending unit
Filter
Oil pump gear
Oil pump

Oil filtration

This is done in two stages. The first takes place at the strainer, located at the bottom of the oil sump. The second stage is the oil filter.

The oil filter is a can-shaped unit attached to the engine block in which an accordion-pleated fibre sheet traps impurities. The degree of filtration of standard quality filters is about 15 to 20 microns. When pressure loss due to a clogged filter exceeds 1 bar/cm², a by-pass valve, located inside the oil filter, opens. Oil is then sent into the main oil line without going through the filter.

Oil pump

The oil pump is one of the engine's most important auxiliary components. Proper lubrication is the key to the engine's integrity, performance and longevity, and it depends on the efficiency of the oil pump. No matter what type of pump it is, pinion or rotary, its operation is identical.

It can be divided into two phases:

Phase 1: intake
The volume of the intake chamber increases; its pressure becomes less than to the pressure in the lower sump; the oil is sucked in.

Phase 2: outflow
The volume of the outflow chamber decreases; pressure increases. The oil pump delivery rate, which is proportional to the engine rpm, is regulated by the pressure release valve. The maximum pressure depends on the engine type. On average it is between 3.5 and 5 bars.

Pinion-type oil pump

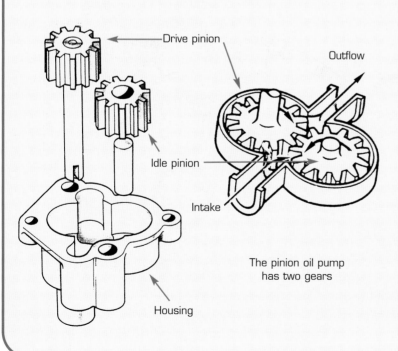

The pinion oil pump has two gears

Perama oil filter

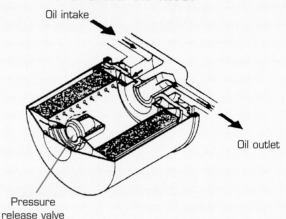

Crescent-type oil pump

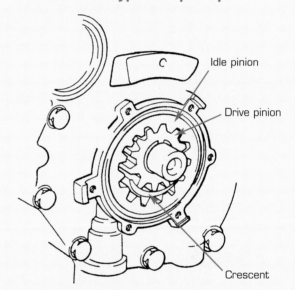

Rotary-type oil pump

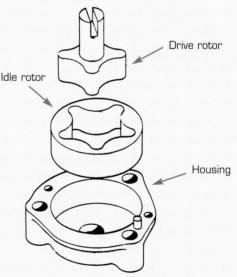

The rotary-type oil pump consists of two offset rotors turning inside a housing with inlet and outlet ports.

OIL TYPES

Do not economise when choosing oil. The difference between a good quality oil and a lesser one is a minimal cost when the health and longevity of your engine are at stake. Which oil do you choose? How do you make the best choice? This section will help you.

The role of oil

Not only does oil reduce friction, and wear of the moving parts, by separating them with a thin film of oil, it plays a role in the tightness of the cylinder seals by stopping micro leaks, especially those in the piston rings. When oil circulates in the engine, it carries away a considerable amount of heat and so contributes to the engine's cooling. It limits corrosion by leaving a protective layer on metal parts. It removes impurities and deposits of carbon and metal particles and other products resulting from incomplete combustion.

So, the role of a lubricant is to:

◆ reduce wear;
◆ reduce energy loss due to friction;
◆ maintain the airtightness of the combustion chamber;
◆ ensure protection against corrosion;
◆ conduct heat away;
◆ clean the engine.

Types of friction that cause mechanical wear

There are three types of friction caused during engine operation:

◆ Dry friction
◆ Oily friction
◆ Fluid friction

Dry friction

This is rare but devastating to the engine. It can happen on starting after a long period of disuse. During the period when the engine wasn't being used, all of the oil has dripped down into the sump; the oil film left is very thin and sometimes nonexistent. On starting the engine, the cams and various bearings are subject to dry friction causing metal abrasion. The oil circulation in the entire engine isn't effective until several seconds after starting.

There is one measure you can take to avoid problems with dry friction: at the start of the new season, squirt a small amount of oil into the air inlet and turn over the engine (without starting it). This will coat the cylinder surfaces with oil and cut down on the friction. Changing the oil frequently will also prolong the life of your engine.

Oily friction

Under heavy load, particularly on starting and during abrupt rpm variations, the protective film of oil coating the metal surfaces can be disrupted. Friction between the rough surfaces of the metal, aggravated by any loose metal particles, causes microscopic metal abrasion and engine wear.

 Oily friction affects gearboxes as well as engines.

Stages of hydrodynamic lubrication of journals*

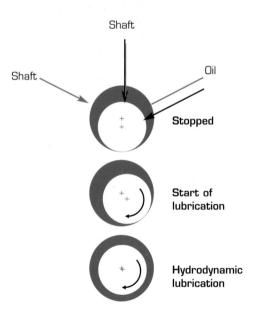

Shaft

Shaft

Oil

Stopped

Start of lubrication

Hydrodynamic lubrication

*The part of the crankshaft that rotates against a bearing.

Fluid friction

This is when a lubricant is introduced between two moving metal parts in such a way that there is no contact between the surfaces (hydrodynamic lubrication, see drawing below).

This is used for the main crankshaft bearings (see diagram left). It is the ideal configuration; no wear should be possible. For this to be true, however, the engine would have to run at a constant rate and never stop.

Conclusion

So we can see that the lubricant is a determining factor. It reduces friction and wear by getting between two moving components to prevent any direct contact.

Contrary to what one would believe, it is not long periods of use that cause the most significant engine wear but the cumulative effect of all the starting and warming up periods. On these occasions, load is placed on the engine while the oil is still cold and circulating poorly.

Moisture condenses in the engine forming droplets in the oil. The water mixes with chemical residues from combustion to form harmful deposits. So you can see how important it is to change the engine oil frequently.

Friction levels

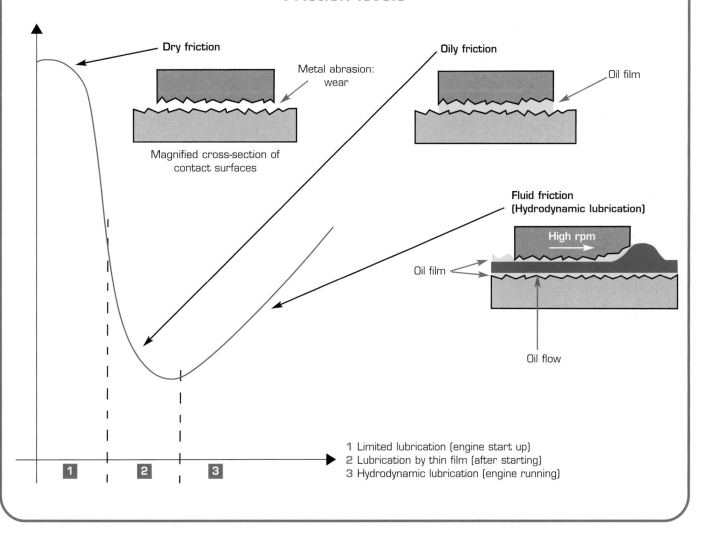

Dry friction

Metal abrasion: wear

Magnified cross-section of contact surfaces

Oily friction

Oil film

Fluid friction (Hydrodynamic lubrication)

High rpm

Oil film

Oil flow

1 Limited lubrication (engine start up)
2 Lubrication by thin film (after starting)
3 Hydrodynamic lubrication (engine running)

Oils generally have the same appearance, texture and odour, but perform differently, depending on their formulae.

API standard

| SA | SB | SC | SD | SE | SF | SG | SH | SJ | SL |

Performance scale – petrol (gasoline) engine

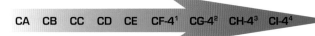

| CA | CB | CC | CD | CE | CF-4[1] | CG-4[2] | CH-4[3] | CI-4[4] |

Performance scale – diesel engine

(1) Replaces indices CD, CE, for normal and turbo compression engines.
(2) Replaces indices CD, CE, CF-4, for engines using fuel containing less than 0.5% sulphur. It has dispersive and anti-wear properties.

Which oil do you choose?

First, check the correct oil viscosity in your engine manual. The SAE (American Society of Automotive Engineers) classify the viscosity of oils in relation to that of water. Tests are carried out under very precise conditions (norm DIN 51 511). Tests at –18°C give the numbers 5 10 15 20 followed by the letter W (winter). The smaller the number, the thinner the oil, making starting easier. The higher the number, the thicker the oil when hot, which gives better resistance to high loads when the engine is hot. This classification only indicates viscosity; it is not a guarantee of quality.

Check the quality

Once you know the viscosity, the quality of oil or the performance criteria can be decided. Price cannot necessarily guide your choice. Here too, there are standards. These are indicated in the small print on the back and sides of the container.

When it comes to quality, several standards define oil performance criteria: the API standard, of American origin, and the ACEA, which is more specific to European engines. Some manufacturers with special requirements have developed oil formulae specific to their engines or to their operating conditions, as is the case with some military units, the manufacturers of VAG, VW, national railways and RATP, etc.

The API standard
The most widespread classification is that of the American Petroleum Institute (API). This standard identifies oils by means of a number of tests, with the help of a code. Example: API SL CH. The letter S (spark) means this type of oil is designed for petrol engines. For diesel engines the letter C (compression) is used. The second letter classifies oils on a scale from A (lowest performance) to L (highest performance) for petrol engines, and from A to I for diesel engines. In the example given, we are looking at an oil that is suitable for use in both petrol (gasoline) and diesel engines.

The classifications SA, AB, SC and SD, and CA, CB, CC, are now obsolete. Note also that there is no 'I' quality in petrol (gasoline) (SI generally means International System).

The European standard
The ACEA standard (European Manufacturers Association) in use since 1996 replaces the former CCMC standard, although its marks are still seen on containers. It takes into account the new pollution emissions standards. Three letters define the engine categories. **A** for petrol engines, **B** for diesel engines and **E** for high capacity industrial engines. Each specification group comprises several performance levels. The number following the category classifies the oil on a scale of 1 to 3 for petrol (gasoline) engines, 1 to 5 for diesel engines and 1 to 3 for industrial engines. The last or last two numbers indicate the year the oil was put on the market. Example: ACEA B3-01 means an oil for a diesel engine, high performance, year 2001.

An oil might be suitable for use in either a petrol (gasoline) or diesel engine, but its use may be classified as optimal for one, and average for the other.

Mineral or synthetic?

Whether mineral or synthetic, engine oil is always derived from crude oil. An oil is said to be mineral when the original product has not been modified. A semi-synthetic oil achieves the same formulation by adding synthetic additives to enhance its performance. Synthetic oil is obtained from crude oil whose molecules are modified by chemical treatments.

At either low or high temperature, synthetic oils offer superior performance, incomparable to that of conventional mineral oils. They have a better viscosity range, great resistance to high temperature and very low volatility. This oil type has been developed to satisfy engine manufacturers' ever increasing demands. Their high price is justified by their exceptional qualities, designed for heavy duty engine use.

Environment

In response to the latest emissions specifications, the ICOMIA 27-92 biodegradability standard stipulates that only oils that are 66% biodegradable can claim to meet the standard.

Economy

A new seal issued by API is given to oils guaranteeing fuel economy.

A seal of approval issued by API is given to oils guaranteeing fuel economy. The symbol indicates the API service class, the viscosity and when appropriate, fuel economy.

CCMC classification

Performance scale – petrol (gasoline) engine

Performance scale – diesel engine

European standard (ACEA)

Oil designed for petrol (gasoline) engines		Oil designed for diesel engines		Oil designed for industrial engines	
A1	Fuel economy.	B1	Minimal use. Fuel economy.	E1	Moderate performance turbo engine.
A2	Normal use. Moderate performance.	B2	Standard quality. Moderate use.	E2	High performance turbo engine.
A3	Severe use. High performance.	B3	Good quality. Severe use.	E3	Very high performance turbo engine. Good stability.
		B4	Direct injection.	E4	Very high performance turbo engine. Fewer oil changes, polishing, dispersion of soot.
A5	High performance. (A1 + A3)	B5	High performance. (B1 + B2 + B3)	E5	Very high performance, high demand turbo engine. Limits wear and deposits in the turbo.

Table of viscosity and oil use

External temperature
in degrees C

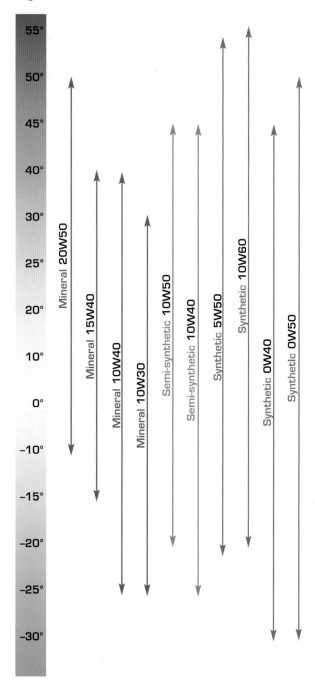

Summary

◆ For your four-stroke engine use a multi-grade oil with the viscosity designed for your operating environment.

◆ Choose an oil based on the engine manufacturer's recommended standards and not on the brand name or advertising pitch.

◆ Numerous manufacturers recommend a certain brand of oil. This is due to commercial agreements. You are free to choose your oil brand. However, you must stick to the quality level recommended by the manufacturer.

◆ You can mix oils of whichever brand you wish without worry – a single grade with a multi-grade; a semi-synthetic with a synthetic. The drawback in this is that the quality obtained will be closer to the lower performance oil.

◆ When buying oil, you will find that prices are generally lower in supermarkets, but don't expect any advice.

◆ Check your oil level regularly. Remember you should top it up before it reaches the minimum level.

◆ Follow the engine manufacturer's recommended schedule of oil changes. Do not unreasonably exceed the number of hours recommended by the manufacturer.

◆ For peace of mind don't go below these oil quality minimums:

B3 Diesel engine
E3 Industrial engine
A3 Petrol (gasoline) engine

◆ Avoid leaving used oil in the engine during storage. At the end of the season, drain the oil (while the engine is still hot) and replace it with new oil.

◆ Use the containers (found in most marinas) for the disposal of used oil.

OIL CONSUMPTION

Need to add oil between oil changes, blue smoke in the exhaust, traces of oil under the engine and in the bilge? Should we worry? At what point are repairs imminent? Which oil should we add?

Is it normal that my engine burns oil?

Yes. An engine consumes oil because of the way it works. You only need to worry if consumption seems to be excessive. I would simply say that adding more than one litre of oil in between oil changes is reason to worry, even if the manufacturer sets the alert threshold at a higher level.

Causes of over-consumption of oil

Seal failure

This can be caused by piston ring and cylinder wear (see diagram below), as seen often in engines with a high number of running hours. But in recreational boat engines used for short periods, the problem is most often caused by clogged piston ring grooves. The rings no longer provide a seal; the oil escapes and burns in the combustion chamber. When a ring groove is clogged, the ring sticks in the groove and doesn't un-stick even under the effects of high temperature. This clogging is caused by oil burning in the groove. The engine continually emits a bluish, smelly smoke out of the exhaust that is particularly dense during acceleration. You can see another sign of this at the engine's breather or by removing the rocker cover oil filler cap. Light bluish puffs of smoke indicate gas leaks coming from the combustion chamber.

Parts of a piston

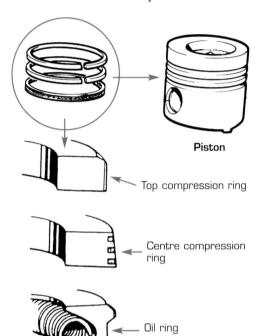

Piston

Top compression ring

Centre compression ring

Oil ring

 The rings, particularly the oil ring, have to fit the sleeve perfectly along the entire stroke to avoid the engine buring oil.

Oil consumption in the piston cylinder

The passage of oil between piston ring and sleeve

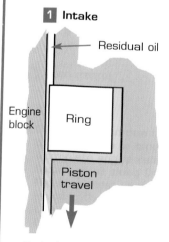

1 Intake

Residual oil

Engine block

Ring

Piston travel

During intake, the ring pushes against the upper face of the groove on the piston and the oil enters the back of the ring.

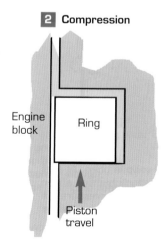

2 Compression

Engine block

Ring

Piston travel

During compression, the oil caught behind the ring finds free upward passage and goes under the next ring.

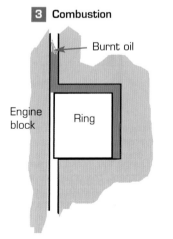

3 Combustion

Burnt oil

Engine block

Ring

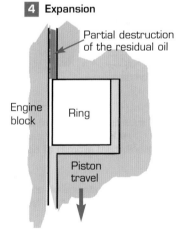

4 Expansion

Partial destruction of the residual oil

Engine block

Ring

Piston travel

During expansion, just after combustion begins, the oil reaches the top compression ring (the first on the piston) and is burnt.

How the gearbox works

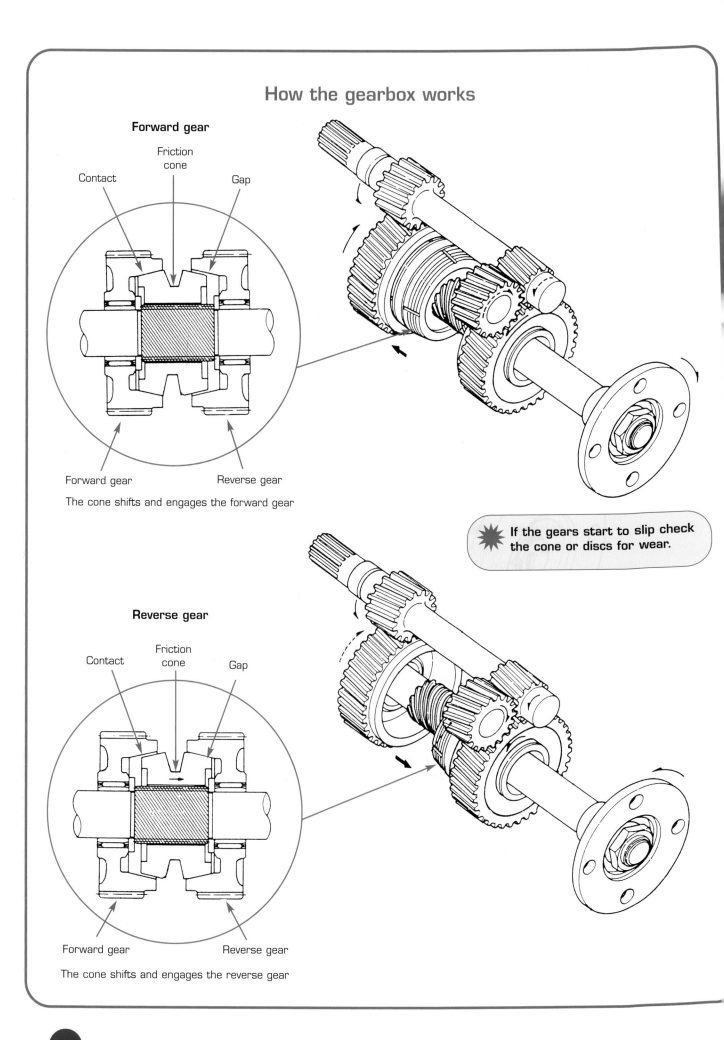

Forward gear

Contact

Friction cone

Gap

Forward gear

Reverse gear

The cone shifts and engages the forward gear

If the gears start to slip check the cone or discs for wear.

Reverse gear

Contact

Friction cone

Gap

Forward gear

Reverse gear

The cone shifts and engages the reverse gear

OIL CONSUMPTION

Need to add oil between oil changes, blue smoke in the exhaust, traces of oil under the engine and in the bilge? Should we worry? At what point are repairs imminent? Which oil should we add?

Is it normal that my engine burns oil?

Yes. An engine consumes oil because of the way it works. You only need to worry if consumption seems to be excessive. I would simply say that adding more than one litre of oil in between oil changes is reason to worry, even if the manufacturer sets the alert threshold at a higher level.

Causes of over-consumption of oil

Seal failure

This can be caused by piston ring and cylinder wear (see diagram below), as seen often in engines with a high number of running hours. But in recreational boat engines used for short periods, the problem is most often caused by clogged piston ring grooves. The rings no longer provide a seal; the oil escapes and burns in the combustion chamber. When a ring groove is clogged, the ring sticks in the groove and doesn't un-stick even under the effects of high temperature. This clogging is caused by oil burning in the groove. The engine continually emits a bluish, smelly smoke out of the exhaust that is particularly dense during acceleration. You can see another sign of this at the engine's breather or by removing the rocker cover oil filler cap. Light bluish puffs of smoke indicate gas leaks coming from the combustion chamber.

Parts of a piston

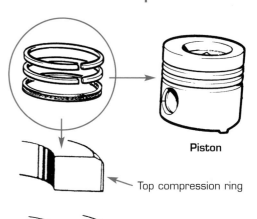

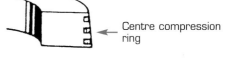
Piston

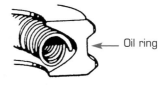
Top compression ring

Centre compression ring

Oil ring

The rings, particularly the oil ring, have to fit the sleeve perfectly along the entire stroke to avoid the engine buring oil.

Oil consumption in the piston cylinder

The passage of oil between piston ring and sleeve

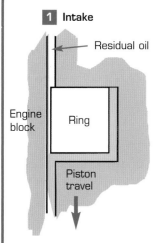

1 Intake

Residual oil

Engine block — Ring — Piston travel

During intake, the ring pushes against the upper face of the groove on the piston and the oil enters the back of the ring.

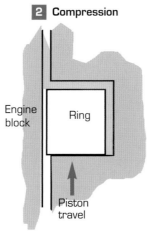

2 Compression

Engine block — Ring — Piston travel

During compression, the oil caught behind the ring finds free upward passage and goes under the next ring.

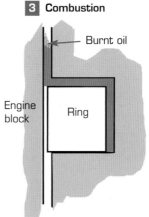

3 Combustion

Burnt oil

Engine block — Ring

During expansion, just after combustion begins, the oil reaches the top compression ring (the first on the piston) and is burnt.

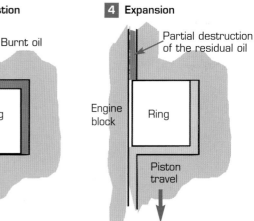

4 Expansion

Partial destruction of the residual oil

Engine block — Ring — Piston travel

Valve stem seal wear

During the intake stroke, oil is drawn in along the valve stems and enters the combustion chamber. The engine emits a bluish, smelly smoke out of the exhaust, particularly when starting. The extent of the problem is related to the level of consumption. Replacing the valve stem seals if necessary often requires removing the cylinder head.

Passage of oil between the valve stem and guide

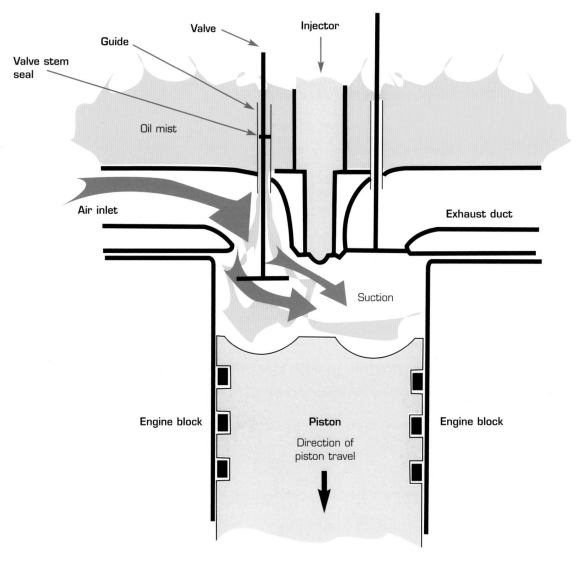

During the intake stroke, the intake valve is open. At that moment, the pressure in the combustion chamber and in the intake manifold is less than atmospheric pressure. Oil is sucked between the guide and the valve stem. If there is significant play in the valve guide assembly, this phenomenon is greatly amplified.

Note: Fitting new valve stem seals eliminates all suction between the valve stem and guide.

Oil leaks

An engine may lose oil because one or several of the seals are no longer airtight. The precise source of the leak must be identified. Risk to the engine is not substantial as long as the leaks are not significant and the oil level is maintained. Should any significant leak appear, check and top up the oil level regularly and never let the engine run without oil.

The amount of work required depends on the type of seals to be replaced. The job is relatively easy if it is a matter of changing the gasket on the rocker cover. In contrast, changing the gasket on the rear main crankshaft bearing necessitates the removal of the engine. Finding the exact source of the leak or leaks is of paramount importance. Thoroughly degrease the engine before looking for the source of the leak or leaks.

Oil consumption through suction, combustion, or external leaks

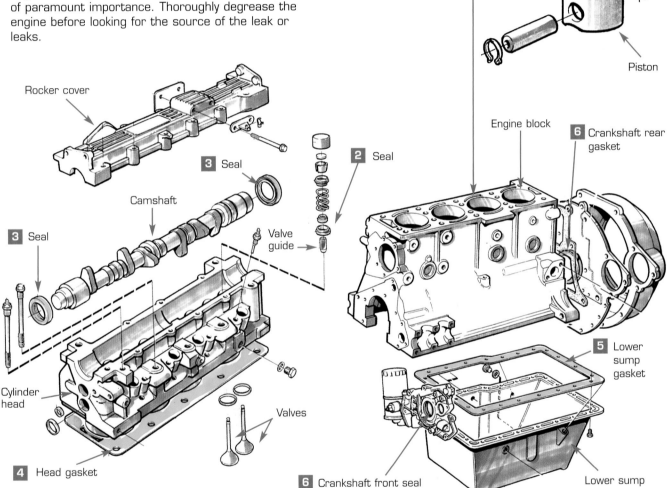

1 Oil consumption due to cylinder and piston ring wear or to sticking due to clogged ring grooves.
2 Suction of oil along the valve stems (intake).
3 Leak or seepage around seals.
3 Leak or seepage around seals.
4 Head gasket
5 Lower sump gasket
6 Crankshaft rear gasket
6 Crankshaft front seal
Lower sump
Rocker cover
Camshaft
Cylinder head
Valves
Valve guide
Seal
Seal
Piston ring
Piston
Engine block

1 **Oil consumption due to cylinder and piston ring wear or to sticking due to clogged ring grooves.**
A complete engine rebuild should be considered.
2 **Suction of oil along the valve stems (intake).**
Replacing these seals on this type of engine requires removing the camshaft.
3 **Leak or seepage around seals.**
Changing this type of seal presents no particular problem. In this case the camshaft pulley must be removed and the timing re-set.
4 **Leak around the head gasket.**
In an internal combustion engine this seal is, without a doubt, the one that needs the most attention, as it is

subject to high pressure, huge and frequent temperature variations, and other types of stress. In the case of a slight leak or seepage, you will need to remove the cylinder head to change the gasket.
5 **Leaks around the sump gasket.**
Changing this gasket, which requires no special skill, will in most cases still require the removal of the engine.
6 **Changing the seals on the sump.**
Changing the front seal poses no particular problem; it is just a matter of removing the pulley that drives the alternator and water pump – unlike the corresponding one in the rear end, which requires removal of the flywheel bell housing and flywheel.

THE ENGINE COOLING SYSTEM

The engine cooling system's function is to:

◆ Disperse the heat in the cylinders generated by combustion.
◆ Maintain the temperature of different components at compatible levels.

Its role is essential to maintain the life and even running of the engine.

Air cooling

The simplest technique consists of blowing a strong flow of air over the cylinders. The advantage of this system is its simplicity – no valves, no pump nor heat exchanger. It involves minimal cost.

But it also has some disadvantages: it makes more noise than a water cooled engine and a very large flow of fresh air is required for cooling. Air cooling, largely used on motorbikes and small engines, is only suitable for totally open boats.

Water cooling

In this system, water is used as the cooling medium and absorbs heat as it flows round the engine block and through the cylinder head. There are two systems of water cooling:

◆ Direct cooling
◆ Indirect cooling.

Direct cooling

This is a simple system where seawater is pumped directly round the engine block and cylinder head – but it has serious disadvantages. To prevent an accumulation of salt and calcium deposits the water temperature must never exceed 40°C. This temperature is too low to ensure thermodynamic efficiency. The result is a noisy engine, early wear, lower fuel economy, increased pollution and risk of corrosion and deterioration of the cooling assembly.

Indirect cooling

With this system, the engine is not directly cooled by seawater but by fresh water, which is itself cooled by seawater by a heat exchanger.

The fresh water system is a closed circuit. The temperature of the water flowing through the cylinder head and around the cylinders is regulated by a thermostat that restricts the flow of water in the exchanger.

Seawater drawn in through a hull fitting located below the waterline is circulated by a displacement pump driven by the engine. After having passed through the exchanger, the heated seawater is expelled into the exhaust manifold.

Direct cooling – seawater circuit

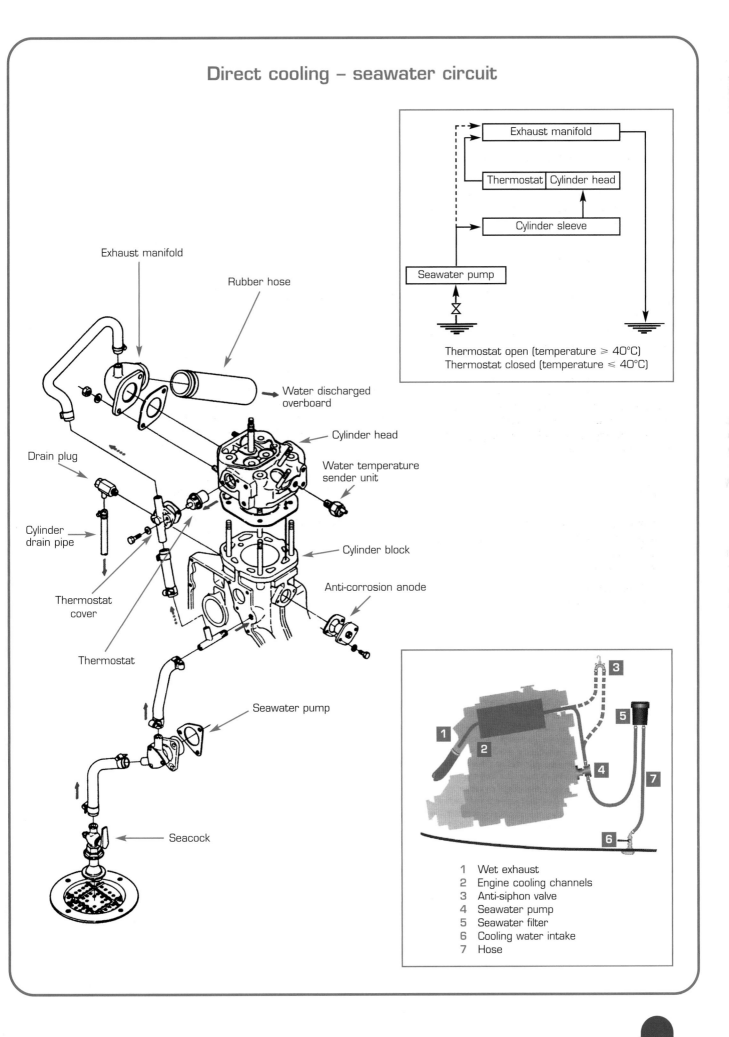

Exhaust manifold

Rubber hose

Water discharged overboard

Cylinder head

Water temperature sender unit

Drain plug

Cylinder drain pipe

Cylinder block

Anti-corrosion anode

Thermostat cover

Thermostat

Seawater pump

Seacock

Exhaust manifold

Thermostat | Cylinder head

Cylinder sleeve

Seawater pump

Thermostat open (temperature ⩾ 40°C)
Thermostat closed (temperature ⩽ 40°C)

1 Wet exhaust
2 Engine cooling channels
3 Anti-siphon valve
4 Seawater pump
5 Seawater filter
6 Cooling water intake
7 Hose

Indirect cooling

Freshwater and seawater circuits

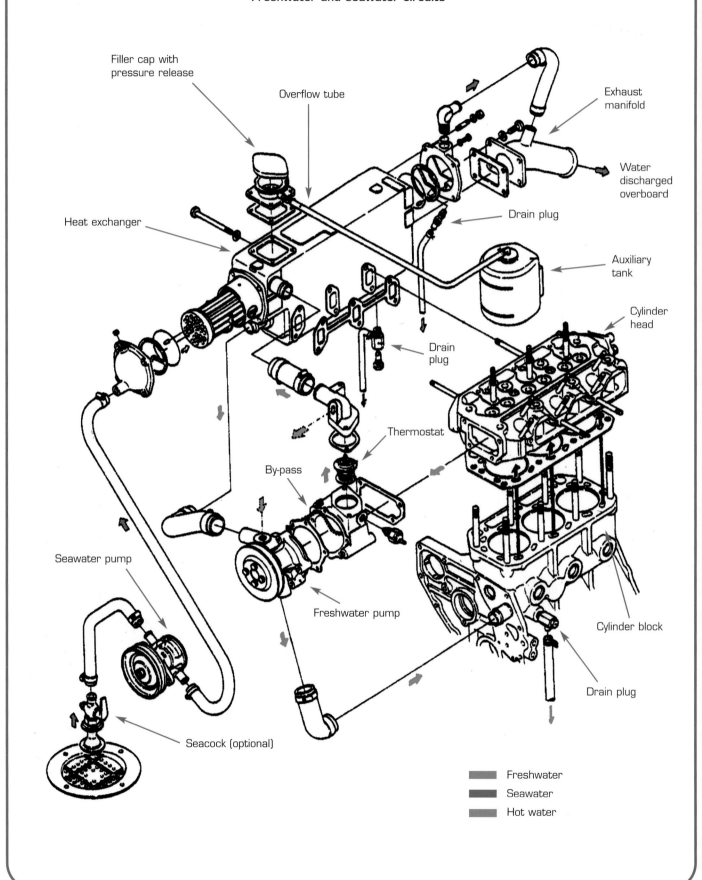

Filler cap with pressure release

Overflow tube

Exhaust manifold

Water discharged overboard

Heat exchanger

Drain plug

Auxiliary tank

Cylinder head

Drain plug

Thermostat

By-pass

Cylinder block

Seawater pump

Freshwater pump

Drain plug

Seacock (optional)

Freshwater

Seawater

Hot water

Indirect cooling

Freshwater and seawater circuits in the heat exchanger

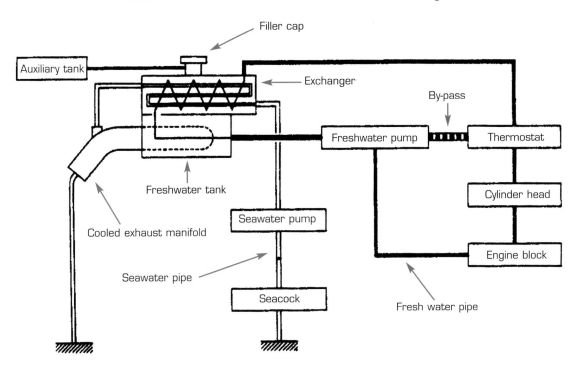

Filler cap

Auxiliary tank

Exchanger

By-pass

Freshwater pump

Thermostat

Cylinder head

Freshwater tank

Engine block

Cooled exhaust manifold

Seawater pump

Seawater pipe

Seacock

Fresh water pipe

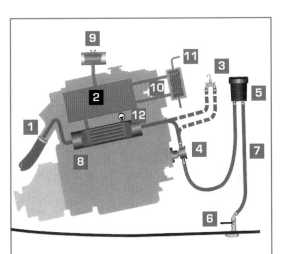

1 Exhaust manifold for wet exhaust systems
2 Engine cooling circuit
3 Anti-siphon valve
4 Seawater pump
5 Seawater filter
6 Cooling water inlet
7 Hose
8 Heat exchanger
9 Expansion tank
10 Hot water outlet
11 Hot water tank
12 Circulation pump

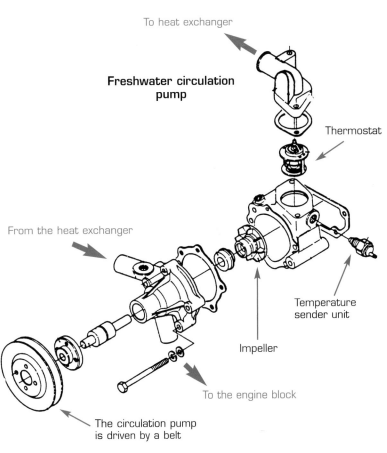

To heat exchanger

Freshwater circulation pump

Thermostat

From the heat exchanger

Temperature sender unit

Impeller

To the engine block

The circulation pump is driven by a belt

System comparisons

	Advantages	Disadvantages
Direct cooling	Simplicity Low manufacturing cost	Weak thermodynamic efficiency (cooling temperature too low) Incomplete combustion (cooling temperature too low) Unavoidable salt deposits Excessive corrosion Risk of freezing (cooling temperature too low) Noisy engine
Indirect cooling	Better thermodynamic efficiency Quieter engine (ideal working temperature of engine 85°–90°C) Lower fuel consumption for equal power No risk of salt deposits or freezing (use of four-season coolant) Possibility of adding a hot water supply or ambient heater into the cooling circuit	More complex system than direct Increases cost of engine The advantages of indirect cooling are such that direct cooling is only used on low power engines of old design.

Exploded view of a seawater pump

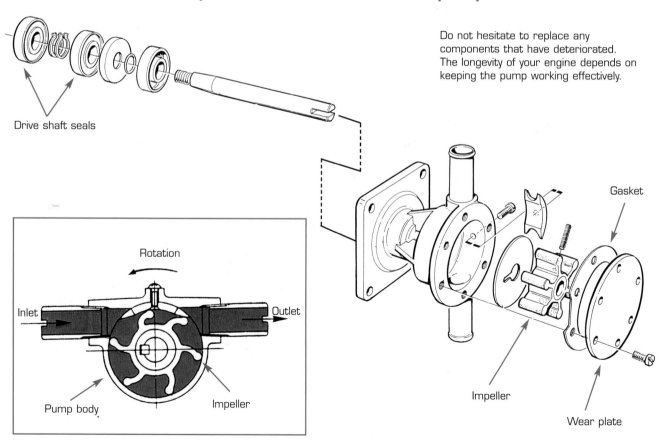

Do not hesitate to replace any
components that have deteriorated.
The longevity of your engine depends on
keeping the pump working effectively.

Drive shaft seals

Gasket

Rotation

Inlet

Outlet

Pump body

Impeller

Impeller

Wear plate

Fresh and saltwater directional flow in the heat exchanger

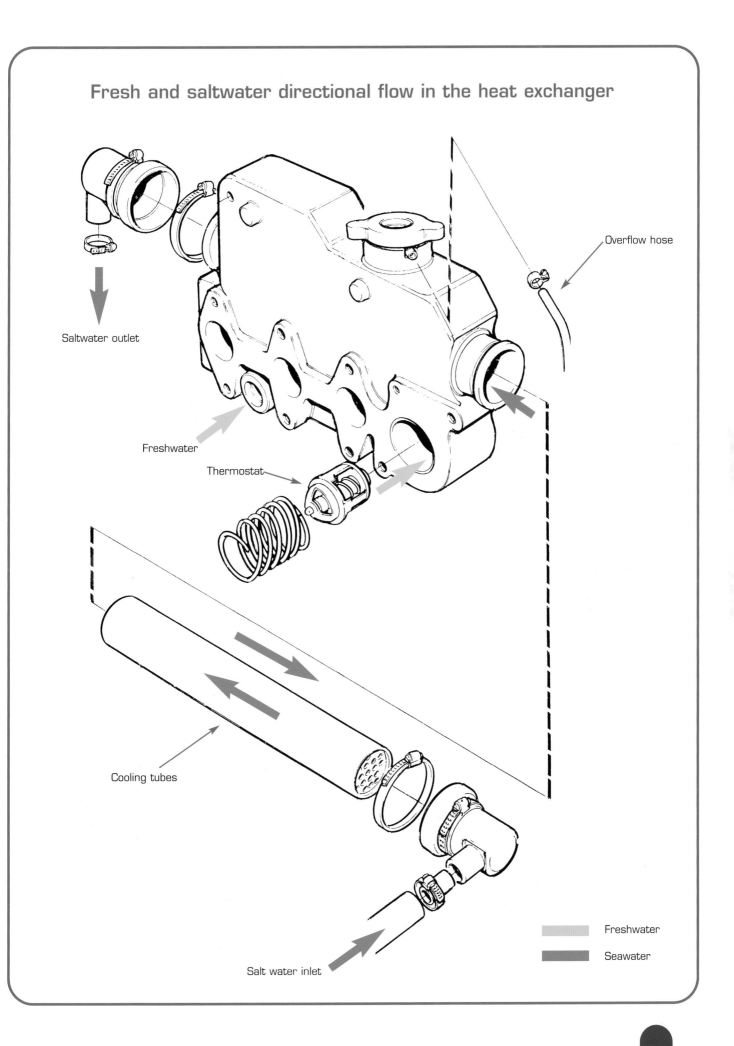

Saltwater outlet

Overflow hose

Freshwater

Thermostat

Cooling tubes

Salt water inlet

Freshwater

Seawater

PROPULSION

To turn the engine's power into thrust, the propulsion unit is composed of four fundamental elements: the engine, the gearbox, the propeller shaft, and the propeller.

The gearbox

Located between the engine and the propeller shaft, the gearbox allows forward and reverse motion of the propeller shaft, and neutral. It also reduces the propeller's rotation ratio to maintain good propeller efficiency. In general, a propeller is much more efficient when it is larger in diameter and rotates more slowly.

Principle

The choice of either mechanical or hydraulic coupling varies between manufacturers and depends on the power and the torque required. The gearbox of an inboard engine is usually composed of three shafts with gears. Forward, neutral, or reverse is selected by shifting a sliding gear.

Schematic view of an inline spur pinion gearbox

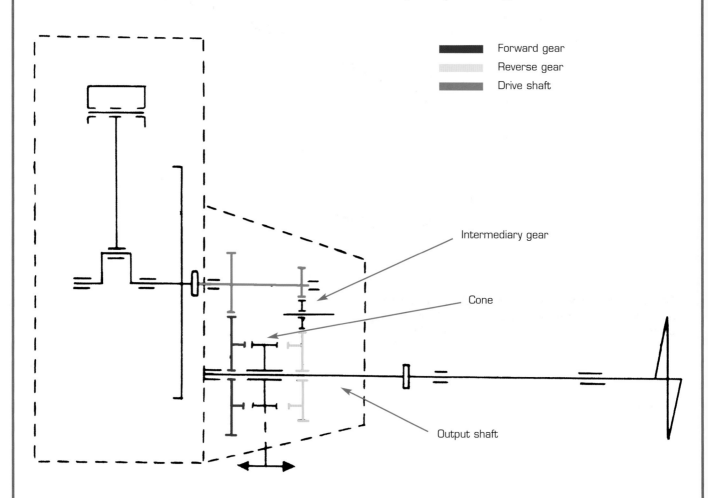

Forward gear
Reverse gear
Drive shaft

Intermediary gear

Cone

Output shaft

Newage PRM 310 gearbox with hydraulic clutch

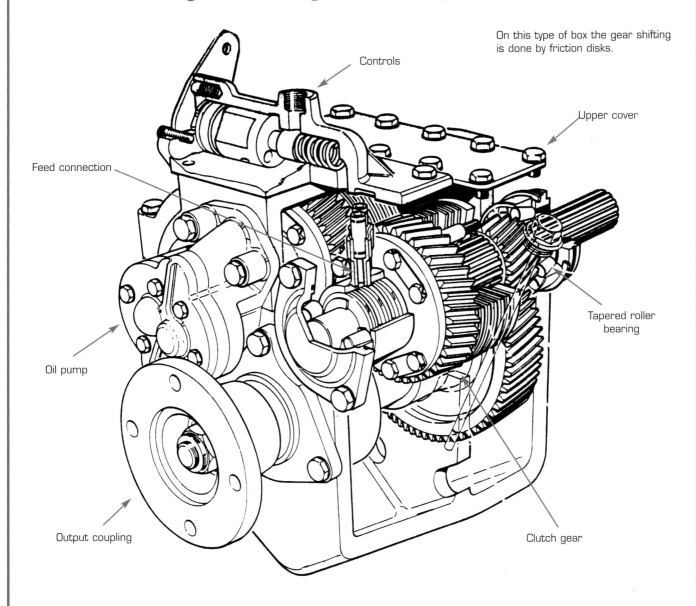

On this type of box the gear shifting is done by friction disks.

Controls

Upper cover

Feed connection

Tapered roller bearing

Oil pump

Output coupling

Clutch gear

The slipping of cone or friction disks allows for smooth meshing of the gears. Splash lubrication of the rotating parts is the most common, simplest, and still the most efficient method. The choice and type of oil and how often it is changed depends on the gearbox technology; it is specified in the manufacturer's maintenance manual for the unit.

This type of simple gearbox is available in several reduction ratios.

The reduction ratio

Each gearbox is defined by its reduction ratio.

Example: a reduction ratio of 2.7/1 means the propeller shaft turns 2.7 times slower than the engine.

How the gearbox works

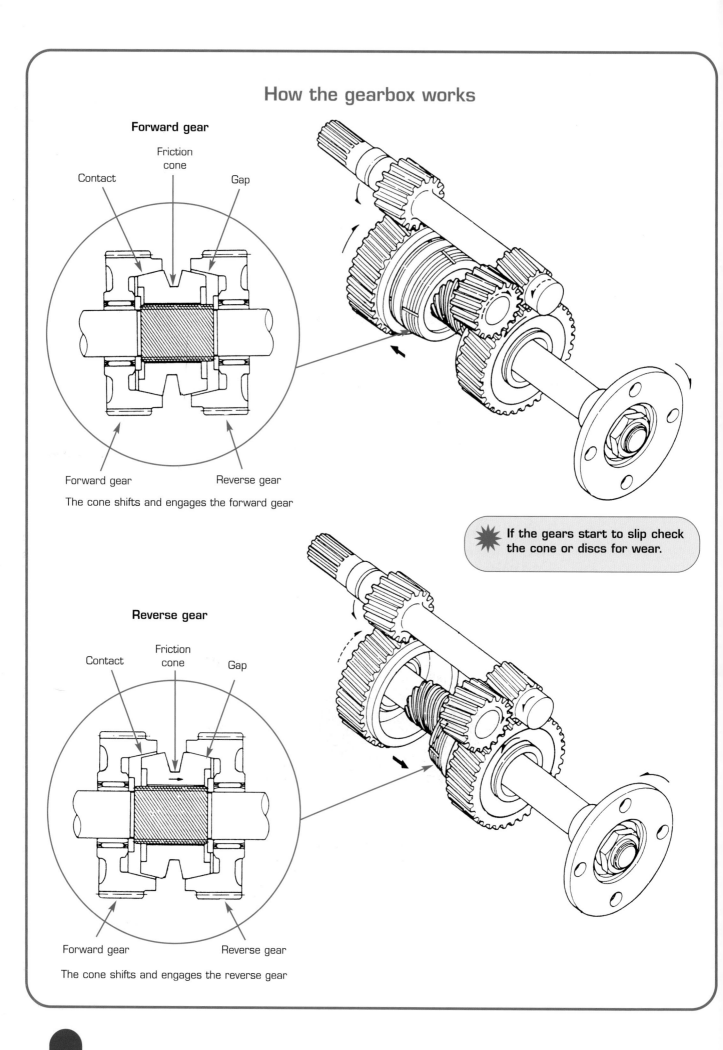

Forward gear

Contact

Friction cone

Gap

Forward gear

Reverse gear

The cone shifts and engages the forward gear

If the gears start to slip check the cone or discs for wear.

Reverse gear

Contact

Friction cone

Gap

Forward gear

Reverse gear

The cone shifts and engages the reverse gear

Bevel gear gearbox

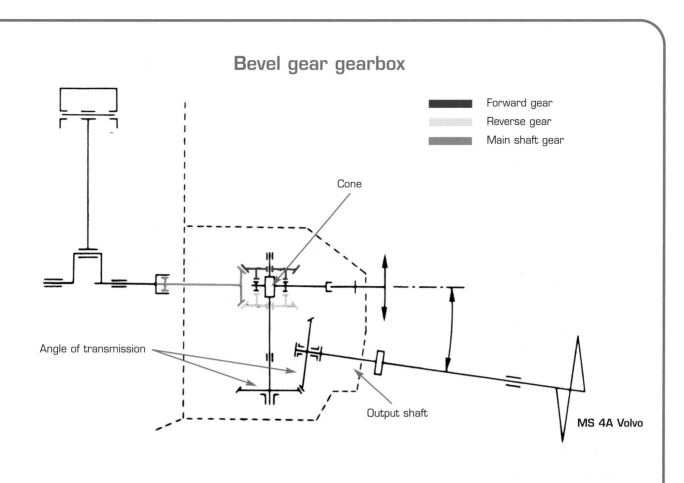

Forward gear
Reverse gear
Main shaft gear

Cone

Angle of transmission

Output shaft

MS 4A Volvo

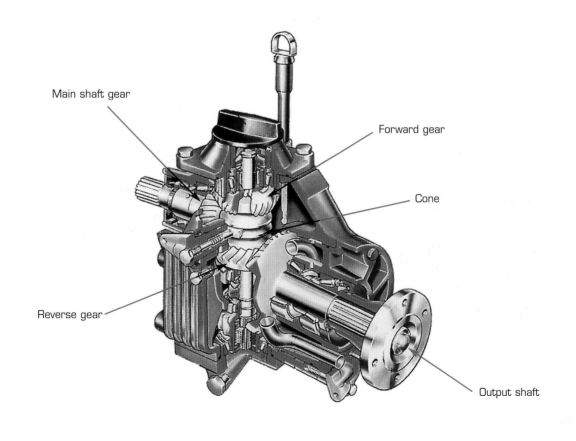

Main shaft gear

Forward gear

Cone

Reverse gear

Output shaft

The drive shaft assembly

There are five parts: the coupling, the shaft seal, the stern tube with its bearings, the shaft, and the propeller. Depending on the type of boat, several assemblies are possible.

Rigid assembly: This type is very common on pleasure craft. The name refers to the rigid mounting of the engine and propeller shaft. It is also very common on fishing trawlers because of its simplicity and robustness, but requires perfect alignment of the shaft and engine. A misalignment would cause the propeller shaft to vibrate, putting too much stress on the bearings and seals.

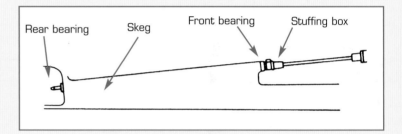

Semi-rigid assembly: This is halfway between a flexible assembly and a rigid assembly.

The stern tube and rear bearing remain but a flexible stuffing box replaces the front bearing. The engine is mounted on rubber blocks to absorb shaft vibrations.

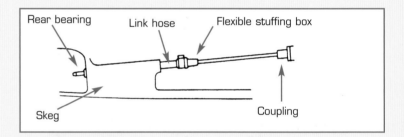

Flexible assembly: This type of assembly provides maximum vibration reduction. The shaft is sealed by a flexible stuffing box. An external strut with a cutless bearing holds the rear end of the shaft in place.

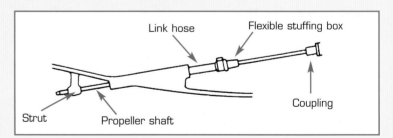

The stuffing box: This is located where the propeller shaft passes through the hull, and is an important part that provides the stern tube seal on the propeller shaft. It was once the source of many problems, now mainly overcome. Although still to be found on some boats, the traditional system of providing a seal by compressing three or four rings of packing around the shaft is disappearing. Manufacturers now use either a rotary or lip seal. These systems are intended to provide a perfect seal and require very little maintenance.

Stuffing box types

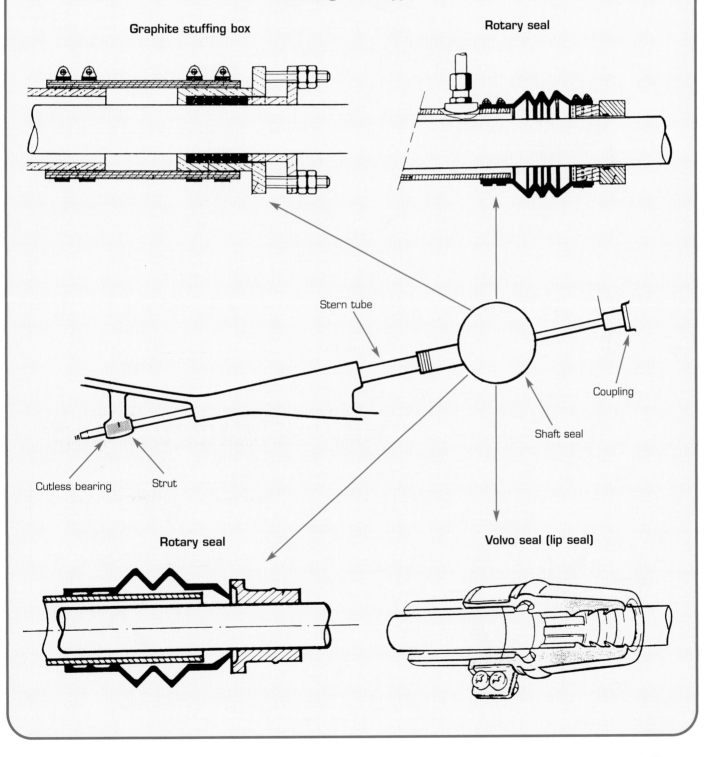

Graphite stuffing box

Rotary seal

Stern tube

Coupling

Shaft seal

Cutless bearing

Strut

Rotary seal

Volvo seal (lip seal)

The propeller: The propulsive force that moves a boat is supplied by the propeller, which is comprised of two or three identical blades around a hub. Several terms describe and characterise the propeller.

The diameter is generally indicated in inches or in centimetres on the propeller hub. For example, in the inscription 13–7, the first numerals indicate the diameter and the second, the pitch. The propeller pitch is the theoretical distance the propeller will travel through the water with one full turn excluding slippage.

Theoretically, a 7 inch pitch means that the propeller moves forward 7 inches at each full turn. In fact, the travel per propeller turn is less due to slippage.

Different propeller types

Twin-blade propeller
Very common on sailboats under 10 metres, it is the best compromise for good performance under power and minimal drag under sail.

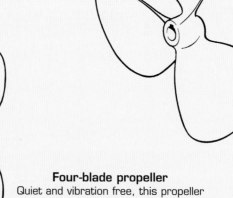

Three-blade propeller
The most efficient for sailboats over 10 metres. The choice of propeller area is a function of boat type, displacement, and desired speed.

Four-blade propeller
Quiet and vibration free, this propeller type is always recommended when a large blade area is required.

Folding, or feathering propeller
Intended to reduce drag while under sail, this type of propeller is very effective but it is less efficient when in reverse compared to a classic fixed two-blade propeller.
Folding propellers are much more expensive than their fixed-blade counterparts.

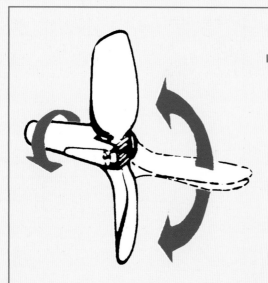

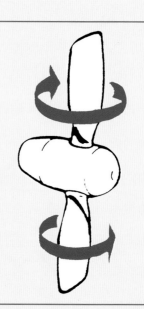

How a propeller works

A propeller is said to act like a screw in the water but in reality it does no such thing because a propeller cannot be driven in water without slipping. A propeller works more like an aerofoil and functions by displacing a certain volume of water which produces a thrust of water from the propeller blades. Without slippage, there is no thrust. On a sailboat, 35 to 45% slippage is acceptable

Two or three blades?

The two-blade propeller is the best compromise for efficiency under power without being a hinderance when under sail. The three-blade propeller increases the drag under sail, but gives greater thrust in forward or reverse and at low revs.

Should the propeller be allowed to spin under sail?

This is a frequently debated question. Theory shows that we should let the propeller spin freely in the water, rather than preventing it from turning by engaging a gear. Most people rarely do. Some fear it might strain the gearbox; others find the spinning propeller too noisy.

Propeller dimensions

Diameter → 16 x 9 ← Pitch

The diameter and pitch marking is usually found on the propeller hub.

Comparative loss of speed under sail

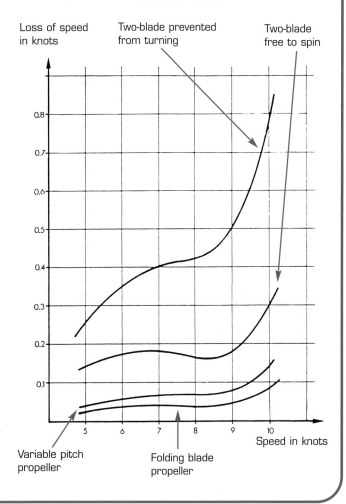

S-Drive transmission

This type of transmission, whose technology is derived directly from the outboard's lower leg, is an integral part of the propulsion unit. It is becoming more and more popular with manufacturers who recognise its many advantages: elimination of the shaft line, ease of assembly, and reduction of vibration and drag. Its disadvantages: the two angle transmissions absorb 20% of the engine's horsepower (an in-line transmission absorbs only 10%).

The transmission has a membrane seal. The leg, made of light alloy, requires the sacrificial anodes to be replaced regularly.

S-Drive transmission schematic

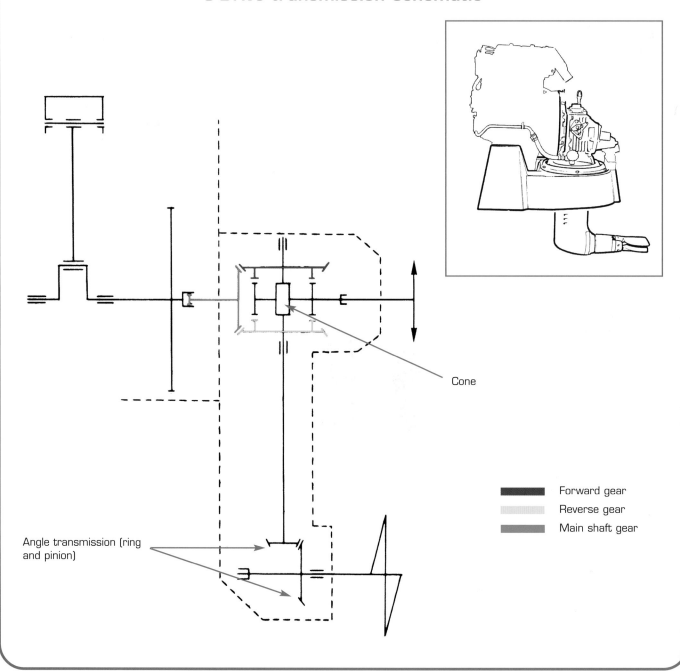

Cone

Forward gear

Reverse gear

Main shaft gear

Angle transmission (ring and pinion)

Cross-section of an S-Drive transmission

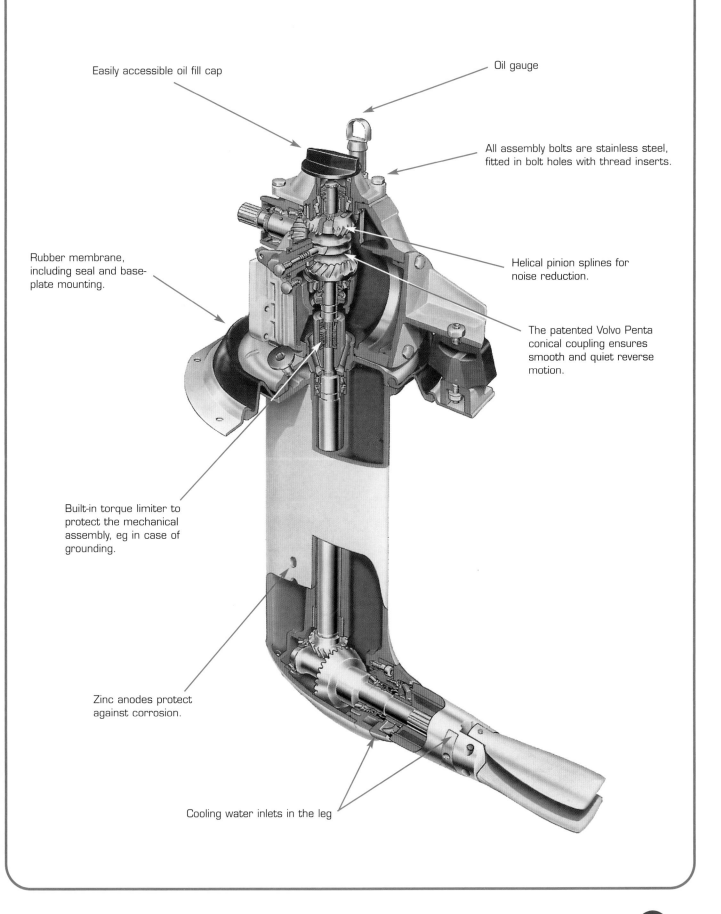

Easily accessible oil fill cap

Oil gauge

All assembly bolts are stainless steel, fitted in bolt holes with thread inserts.

Rubber membrane, including seal and base-plate mounting.

Helical pinion splines for noise reduction.

The patented Volvo Penta conical coupling ensures smooth and quiet reverse motion.

Built-in torque limiter to protect the mechanical assembly, eg in case of grounding.

Zinc anodes protect against corrosion.

Cooling water inlets in the leg

POWER

How much power do I need for my boat? Good question. In fact, in the case of a sailboat, the choice is based on the retailer's advice or on your own judgement.

Other factors such as the space required for the unit, its weight, technology, and its price, all play a role in the final decision.

Engine power

To define the propulsive force needed to push a boat along it is necessary to define exactly what horsepower we are looking at. Engines designed for recreational boating are 'light duty' type. Those designed for professional use are said to be 'heavy duty'. Light duty is defined as less than 200 hours of use per year, generally with short periods of use at full power, followed by periods of use at a lower cruising speed. Anything over that corresponds to 'heavy duty'.

It is also important to distinguish between gross engine power (power at the flywheel) and power produced at the propeller. Discrepancies of 20 to 30% may appear, depending on the type of transmission and number of accessories it drives (water pump, alternator).

Maximum rpm

Engines produce their maximum power at different rpm. This is mainly due to the technologies adopted by the manufacturers. As a general rule, modern engines endowed with the latest technology produce their maximum horsepower at much higher rpm than engines of older design.

For many cruisers, getting the maximum rpm at the tachometer is an obsession. In practice, the maximum acceleration rate will vary depending on the choice of propeller. Too large a propeller (blade or pitch) will absorb almost all of the engine's power before it reaches maximum rpm and the converse is true if the propeller is too small. The cleanliness of the propeller and hull, and the boat's displacement also has a bearing on the quest for maximum rpm.

Estimated horsepower

To simplify, we can say that the engine's horsepower is a function of the boat speed we wish to obtain, which is influenced by waterline length, displacement, and hull shape.

Note the stern wave. The boat has reached its maximum speed.

What top speed do I need?

The waterline length serves as a basis for determining the maximum speed. To calculate this, the following formula is used:

$$\text{Speed} = \sqrt{\frac{\text{Waterline length} \times g}{2 \times \pi}} \times 1.944$$

in which g equals 9.81 metres/second, pi equals 3.1416 and the waterline length in metres.

   Example: a boat with a 12 metre waterline length will be able to reach a displacement speed of 8.41 knots.

$$S = \sqrt{\frac{12 \times 9.81}{2 \times 3.1416}} \times 1.944 = 8.41 \text{ knots}$$

What horsepower?

With this calculation we can estimate the effective horsepower needed to reach a given speed.

$$EH = \frac{S \text{ (in km/h)}}{\sqrt{\text{Waterline length} \text{ (in metres)}}}$$

P = EH × Δ
Δ = displacement in tons
EH = coefficient of effective horsepower
P = minimum required power in hp

Displacement is the basis for determining the horsepower required for the boat to reach the desired speed.

   The definitive choice of horsepower can then be increased to take into account desired course, boat use, operating conditions, safety, power reserve or cost. At a practical level, the solution is to opt for an engine with 20% more horsepower than that which is needed to reach hull speed in flat water without wind.

   It is obvious that the problem is complex and we must take into account these various factors.

Diagram of an engine's electrical circuit

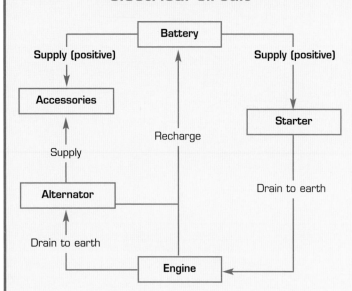

Electricity plays an important role on a boat. The current trend is an increase in electronic accessories. The proliferation of electronic nav-aids and domestic appliances increases the demand for power. While it is often possible to plug into shore power while docked, this is not the case when under sail or at anchor. The onboard electrical system must then be capable of supplying sufficient electricity for your equipment.

Engines are supplied with a standard single-pole electrical system where the engine acts as the earth. A double-pole system is another option. It is usually used on engines in steel hulls and especially in aluminium hulls.

Safety requires the use of two sets of batteries, one for engine starting, one for onboard equipment and appliances.

Principal components of your engine's electrical system

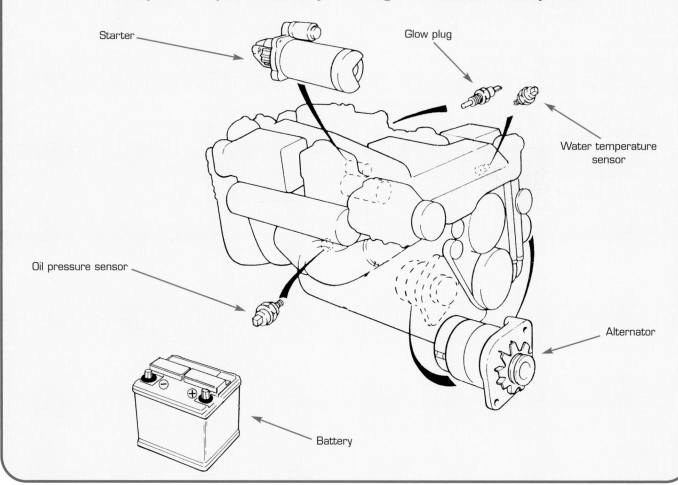

Wiring diagram – single-pole assembly (earthed to engine block)

1 Starter
2 Alternator
3 Oil pressure sensor
4 Oil temperature sensor
5 Water temperature probe
6 Glow plug with its solenoid

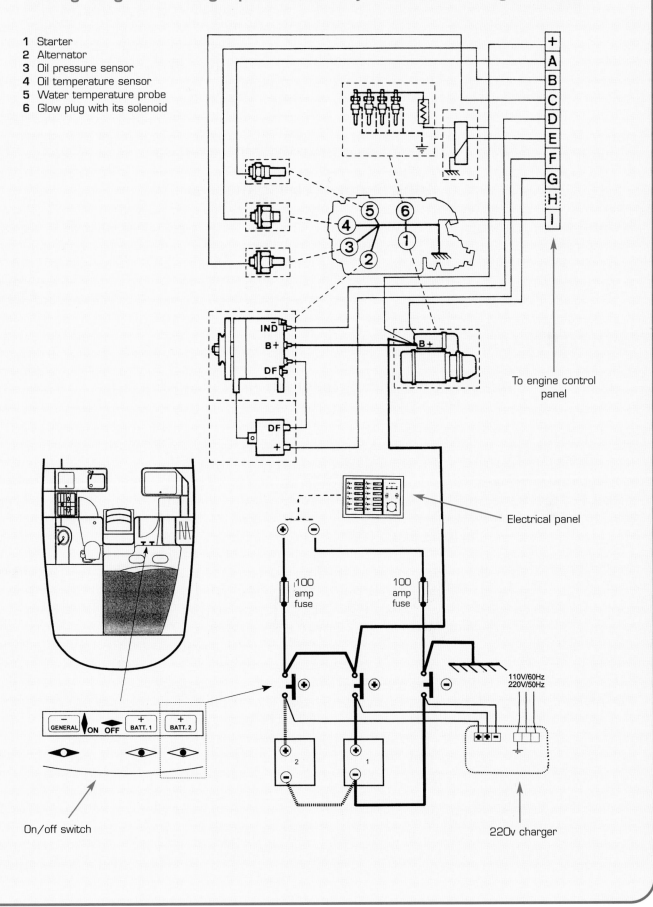

To engine control panel

Electrical panel

100 amp fuse

100 amp fuse

GENERAL ON OFF BATT. 1 BATT. 2

On/off switch

110V/60Hz
220V/50Hz

220v charger

The electrical system consists of:

◆ an electrical potential made up of one or more batteries;
◆ a charging circuit, comprising an alternator and a regulator that will recharge the battery and supply electricity to equipment when the engine is running;
◆ a system to start the engine.

Current legislation requires that the battery be capable of performing a minimum of six consecutive starts without recharging.

Control panel wiring diagram

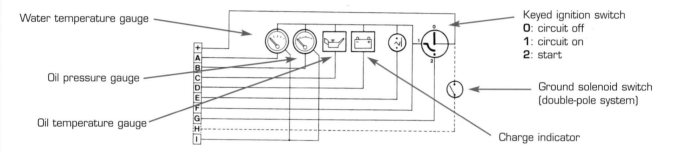

Water temperature gauge

Oil pressure gauge

Oil temperature gauge

Keyed ignition switch
0: circuit off
1: circuit on
2: start

Ground solenoid switch (double-pole system)

Charge indicator

Double-pole wiring diagram (insulated earth)

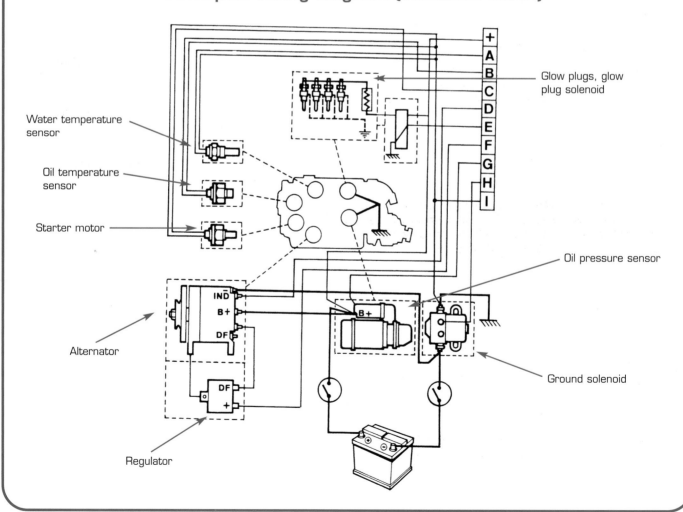

Glow plugs, glow plug solenoid

Water temperature sensor

Oil temperature sensor

Starter motor

Oil pressure sensor

Alternator

Ground solenoid

Regulator

Batteries

Most boats are equipped with lead-acid batteries. The number of lead plates determines the battery's voltage. There are three plates in a 6 volt battery, and six plates in a 12 volt. Each plate is immersed in a solution, the electrolyte, composed of demineralised water (60%) and sulphuric acid (40%). The electrolyte density varies depending on the charge level of the battery.

Pleasure boats are equipped with 12 volt batteries. Series connection allows a 24 volt system for larger installations.

Battery specifications

Battery specifications are marked on the battery top; example: 12 Volts, 80Ah 200A.

Nominal voltage

Expressed in volts, this determines the nominal voltage of the battery; in the above example: 12 volts.

Capacity

The capacity of a battery (Q) is expressed in amp-hours (Ah). In general, the capacity given is for a 20 hour period. It depends on how active the battery is. An 80Ah battery can supply 80 amps in 20 hours or 4 amps per hour over 20 hours.

As an example, a 24 watt bulb connected to this battery will draw:

24 watts = 12 volts x 1 amp, or 24 ÷ 12 = 2 amps. This bulb will drain the battery in 80 ÷ 2 = 40 hours.

A starter motor which draws 300 amps will drain the battery in 80 ÷ 300 = 0.26 hour, that is, around 15 minutes. This is why you need a dedicated starting battery.

The cold current test

This is used to evaluate starting ability at low temperature.

In the example above: 200A is the intensity the battery can supply at −18°C without an element's voltage falling below 1.5 volts after 30 seconds of use.

Choice of battery

First, work out your electrical needs.
Example:

◆ Onboard lighting : 5 x 20 watt bulbs for 6 hours needs 600 watts
◆ Radio: on average, about 100 watts for 6 hours needs 600 watts
◆ Miscellaneous equipment: bilge pump, water pump, etc, approximately 125 watts.

Total consumption: 1325 watts. If the electrical circuit is 12 volts, the consumption will be around 1325 ÷12 = 110Ah.

AC Delco battery

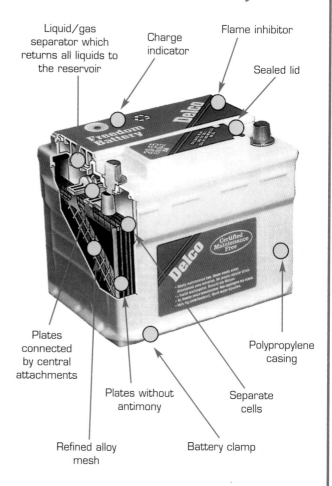

Liquid/gas separator which returns all liquids to the reservoir

Charge indicator

Flame inhibitor

Sealed lid

Plates connected by central attachments

Plates without antimony

Separate cells

Polypropylene casing

Refined alloy mesh

Battery clamp

Battery assemblies

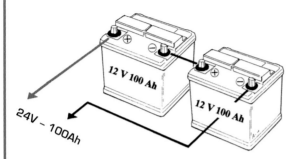

Two 12 volt – 100Ah batteries connected in series doubles their voltage at equal capacity.

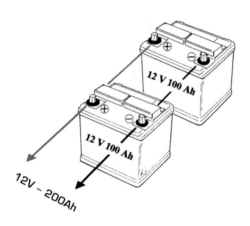

Connecting these same batteries in parallel, their capacity doubles at equal voltage.

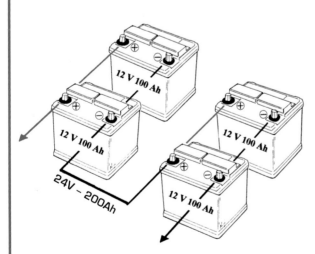

Combining groups in series and in parallel will increase voltage and capacity.

A battery must not be discharged by more than 80% of its capacity, but it is a simple matter to check this. In the above example, the battery capacity should not be under 110Ah
110 ÷ 0.8 = 137Ah.

For safety and comfort, the total battery capacity should be at least equal to double if not triple the daily requirement to avoid deep discharge and excessively long charging time.

> ✸ **This capacity can be divided between two batteries.**

The power of the starter motor also affects the choice of engine battery. Each battery's manufacturer indicates the maximum allowed. Before choosing a battery, check that the maximum starting power drain is sufficient for the motor.

Design

Each manufacturer produces a range of batteries dedicated to marine use.

We distinguish between the traditional lead-acid batteries and maintenance-free batteries. The latter are often 25 to 30% more expensive.

For a boat, choose maintenance-free batteries for performance (low rate of auto-discharge), ease of maintenance and safety.

Battery installations

The simplest system is to allocate one battery for engine starting and one or more batteries for the other onboard needs. It is desirable to be able to charge the batteries simultaneously or independently, and use either one or the other, and even to connect them together to increase their power when needed.

Of course, when replacing one battery with another you need to take into account their characteristics, sizes and the location of their terminals.

The wiring is said to be 'in parallel' when all the batteries' positive and negative poles are grouped separately. In this type of assembly, the output voltage remains the same and equivalent to that of one battery. The available power is the sum of the total number of batteries.

The wiring is said to be 'in series' when the positive pole of one battery is connected to the negative pole of the next. The output voltage is equal to the sum of the batteries' voltages. The capacity remains the same as for one battery.

Replacing one battery with another poses no particular problems. However, if we were to install several batteries or perform repairs on the engine, it would be good to observe these few rules:

- ◆ *Never* install two batteries of different capacities in series.
- ◆ *Never* install two batteries of different voltages in parallel.
- ◆ *Never* invert the positive and negative terminals when assembling.
- ◆ Systematically isolate the batteries (turn off the main switch) when working on the engine or the electrical circuit.

Wiring with a load balancing solenoid or diode

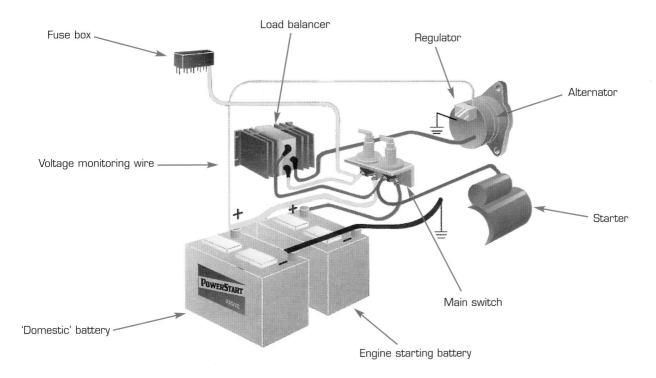

Fuse box

Load balancer

Regulator

Alternator

Voltage monitoring wire

Starter

Main switch

'Domestic' battery

POWERSTART

Engine starting battery

Wiring with a second alternator

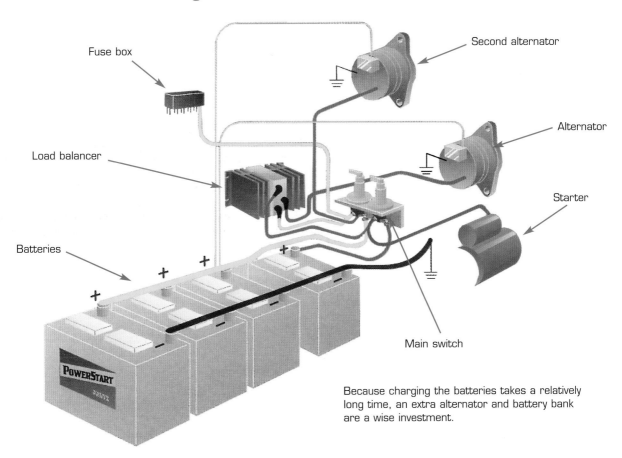

Fuse box

Second alternator

Alternator

Load balancer

Starter

Batteries

Main switch

POWERSTART

Because charging the batteries takes a relatively long time, an extra alternator and battery bank are a wise investment.

Pre-engaged starter with reduction

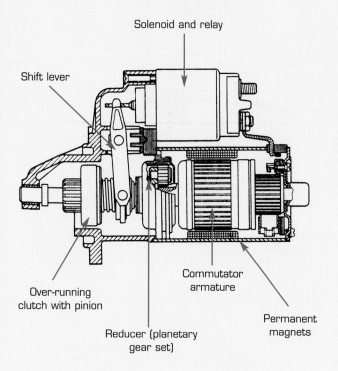

Solenoid and relay

Shift lever

Over-running clutch with pinion

Reducer (planetary gear set)

Commutator armature

Permanent magnets

To start the engine it has to be made to turn over, overcoming the resistance generated by compression and friction. To that end, we have a high-powered electric motor directly engaged on the flywheel. The starter motor axis is extended by a pinion. The flywheel has a crown gear.

It is important to note that resistance is much greater when the engine is cold than when it is hot.

Starter motor structure and operation

The starter comprises:

◆ a direct-current electric motor
◆ a starter
◆ an electromagnetic control

The starter requires a flawless supply of electricity. An insufficient charge, bad battery terminal contacts, dirty or worn brushes are enough to affect starter function.
The starter makes the starter pinion engage onto the flywheel's crown gear, then disengages it once the engine turns. Whichever type: inertia or pre-engaged, the starter's job is very demanding, therefore it is prone to wear.

Inertia-type starter

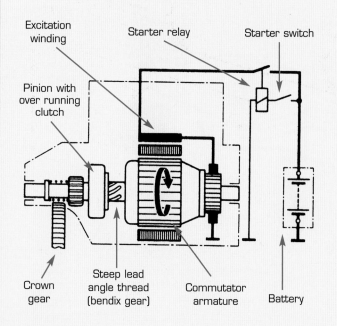

Excitation winding

Starter relay

Starter switch

Pinion with over running clutch

Crown gear

Steep lead angle thread (bendix gear)

Commutator armature

Battery

Electro-mechanical positive drive starter

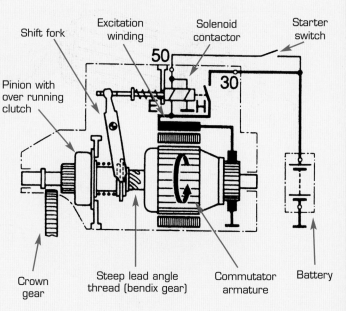

Shift fork

Excitation winding

Solenoid contactor

Starter switch

Pinion with over running clutch

50

30

Crown gear

Steep lead angle thread (bendix gear)

Commutator armature

Battery

Electromagnetic control

Control consists of an electromagnet (a solenoid) located on top of the starter and has the function of pushing the launcher towards the crown gear on the flywheel. The electromagnet has two coils: one coil for attraction and one to sustain contact. In the first phase, both coils act jointly to shift the launcher. Then once the starter turns, in the second phase, the attraction coil is short-circuited. The launcher is retained only by the 'sustain' coil.

Exploded view of a positive drive pre-engaged starter and reducer

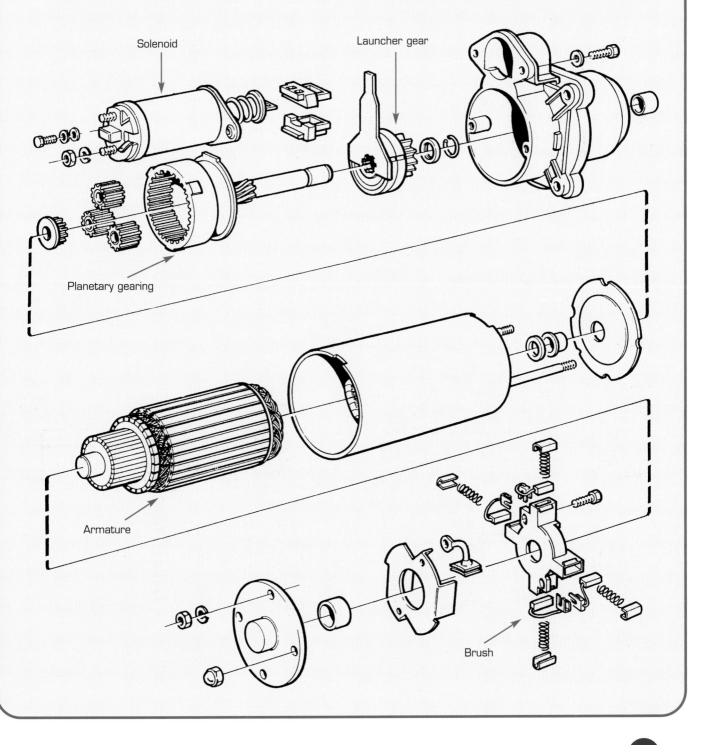

Solenoid

Launcher gear

Planetary gearing

Armature

Brush

Charging system

The original alternator supplied with the engine provides the simplest way to produce electricity. These low-power alternators are designed to fill the energy requirements of a minimum level of equipment.

Alternator structure and operation

The alternator is often driven by a belt connected to a pulley on the crankshaft. If this belt breaks, the alternator stops and the charge indicator light comes on.

The alternator operation, which is based on the principle of electromagnetic induction, comprises a rotor, a stator, a rectifier cell and two brushes. The rotation of the rotor, fed by an excitation current sent to the rotor by two brushes sliding over ring tracks, generates an alternate current in the stator. This is then rectified when it goes through the rectifier cell.

A cutaway view of an alternator

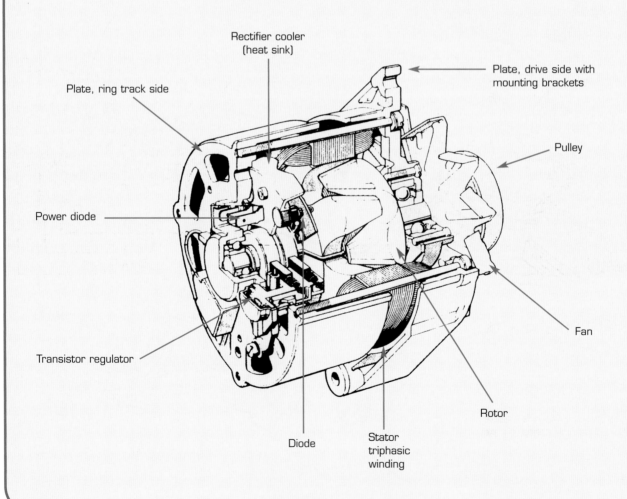

Rectifier cooler
(heat sink)

Plate, drive side with
mounting brackets

Plate, ring track side

Pulley

Power diode

Fan

Transistor regulator

Rotor

Diode

Stator
triphasic
winding

To supply the output rate in accordance with the charge demanded and maintain a constant voltage output, a regulator, integrated or separated from the alternator, modifies the excitation current from the rotor.

> **The regulator limits the battery charge to around 1/10th of its capacity. To charge a 100Ah battery 50% discharged, the engine will have to run for a minimum of five hours for it to reach its nominal capacity. If there is a large bank of batteries onboard and high electricity use, a supplementary means of satisfying demand will have to be found which will permit the batteries to be recharged as quickly as possible.**

To monitor the proper functioning of the charging circuit, engine control panels are equipped with a charge indicator light and sometimes a voltmeter.

Alternator output as a function of engine rpm (for constant voltage)

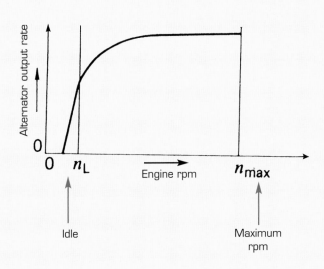

Alternator adjustment

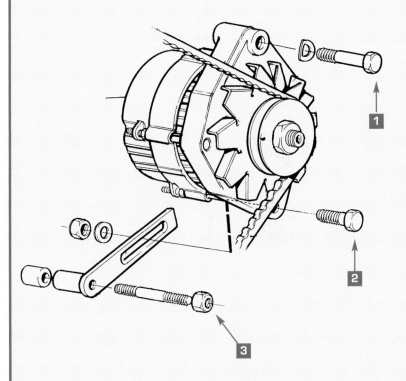

The alternator belt tension is adjusted by pivoting the alternator on its fastening axes 1, 2, 3

Circuit of a triphase alternator equipped with built-in regulator

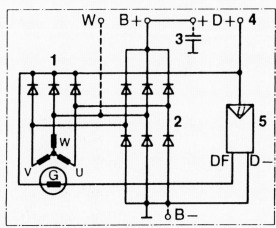

1 Excitation diodes
2 Power diodes
3 Radio interference suppressor
4 To charge indicator light
5 Regulator

MAINTENANCE

JUST LIKE ANY MACHINERY, for a diesel engine to stay in good working order and have a long working life it requires regular maintenance and replacement of the parts that are prone to wear. You will find 25 worksheets in this chapter that will take you through all the maintenance procedures, from changing the oil to adjusting the valves. Step-by-step instructions are given for each task together with clear photographs and captions. The level of difficulty of the job, the time required and tools needed are also indicated.

An engine comprises numerous parts and mechanical units. These work under friction and are subjected to severe pressure and temperatures.

An engine's lifespan depends, of course, on the quality of its design and production, but also on the care its owner is willing to dedicate to it.

Replacement of some worn parts such as belts or consumables like engine and gearbox oil is predictable. Follow your engine's maintenance schedule as recommended in the manufacturer's manual.

Some manuals supplied with the engine are very detailed; others are very brief and not very helpful. If you do not have an engine manual, in this chapter you will find a table indicating what to check, when service should be performed as well as worksheets for the various maintenance tasks. You will, however, need to adapt the operations described in the table to the specific characteristics of your engine.

IMPORTANT

Your inboard engine is designed to work with its original parts. Replace with genuine parts to ensure proper function and fulfil the manufacturer's warranty. Also, follow the schedule for inspection and maintenance closely. Your safety and the longevity of your engine depend on it.

Warranty

When you purchase a boat from the manufacturer or retailer, you are asked to have the scheduled maintenance done by one of the manufacturer's agents or retailers. This is one of the primary conditions for the warranty. Remember to have your maintenance logbook stamped when you have this work done.

IMPORTANT

Having your engine's service done by the manufacturer's agent constitutes an extra warranty to preserve your rights in case of hidden defects. Often, repair of a defect listed by a manufacturer's after sales service is covered even after the warranty's expiry date.

If you decide to carry out some basic maintenance for your engine, you will first need to invest in a good tool set. You will also have to spend some time studying the various systems (transmission, cooling, fuel) on which you are going to work.

If you are new to it or if you have any doubts, then before taking any action, it would be a good idea to review some of the basic theoretical information given at the beginning of this book.

Tools

A rule to remember: always chose good quality tools, obviously more costly, but much more reliable. Remember the saying: 'A good tradesman uses good tools'.

This book does not pretend to replace the proper workshop manual for each engine. However, the maintenance and repair worksheets will guide you and inform you of the general methods and precautions that apply to each job.

General recommendations

Keep the engine and its compartment perfectly clean. In fact, it is by cleaning the engine that we become aware of line ruptures, loose bolts, various leaks, etc, so we can prevent any damage.

During any maintenance or repair, take care not to disassemble or misadjust any factory-sealed unit. These units can only be repaired by the manufacturer, or after the warranty period, by a specialised professional.

When you disassemble any unit in the engine, do it methodically without rushing. Make sure you remember the order of reassembly, especially the different nuts and bolts. When reassembling, ensure that all the parts and surfaces are really clean.

Checking the drive belt tension.

Job	Frequency					
	Daily	Every 50 hours	Every 100 hours	Every 200 hrs/ Yearly	Every 500 hours	Every 1000 hours
Engine						
Check oil level	●					
Change engine oil			■			
Replace oil filter			■			
Clean the engine compartment			●			
Check and possibly adjust tappets				●		
Clean air filter			●			
Check exhaust colour	●					
Cooling						
Check water level	●					
Check operating temperature	●					
Clean seawater filter		●				
Drain the water circuit (heat exchanger)					■	
Check anodes			●			
Check seawater pump impeller				●		
Recondition seawater pump					■	
Check thermostat				●		
Clean heat exchanger					●	
Fuel						
Check fuel level	●					
Replace filter components				■		
Check and adjust injectors					●	
Adjust injection timing					●	
Electricity						
Check battery charge state (control panel)	●					
Check belt tension		●				
Replace belt					■	
Check electrolyte level in the battery		●				
Check connections				●		
Replace brushes on alternator						■
Replace brushes on starter motor						■
Gearbox/reducer						
Check oil level			●			
Change gearbox oil					■	
Drive assembly/unit						
Check stuffing box			●			
Check cutless bearing				●		
Check propeller shaft alignment						●
Check exhaust line (clamps, hose, muffler)				●		
Check remote controls, lubrication, adjustment				●		
Check tightening of engine mount bolts				●		

● Check, adjust if needed   ■ Replace

BASIC TOOL KIT

A leisure boater should always have a minimum set of tools to be able to do simple jobs.

Basic tool kit

Spanners: flat, mixed, or sockets.

 REMEMBER
When working on an engine made in the USA or in England, use spanners with dimensions in inches and fractions of inches.

Screwdrivers: at least two flat screwdrivers and two Philips screwdrivers.
Pliers: general purpose, cutting or vice (mole) grip – each has a specific use.
Hammer and mallet: brutal, but indispensable.

Specific tools

Set of sockets, a feeler gauge and Allen keys: essential, as is a wire brush.
Multimeter: needed in many cases to check the circuits.
Torque wrench: indispensable for following tightening specifications.

Useful products and accessories

Have a **spray can of waterproofing agent** (such as WD 40), some **marine grease**, an **oil can** and some **sealing adhesive tape** (self vulcanising).
Maintenance being quite a dirty job, you had better have some **rags** and **liquid soap** to clean your hands.
Good luck!

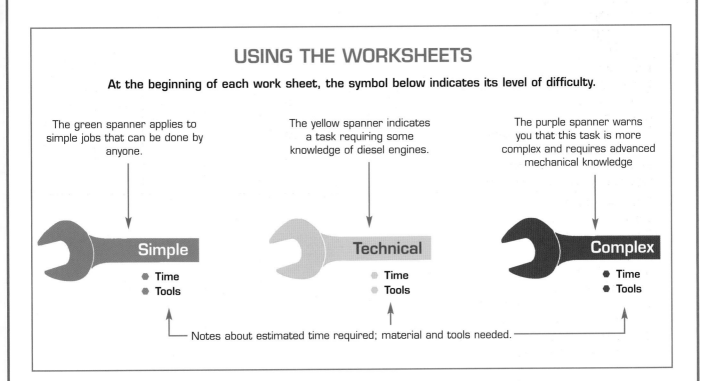

USING THE WORKSHEETS

At the beginning of each work sheet, the symbol below indicates its level of difficulty.

The green spanner applies to simple jobs that can be done by anyone.

The yellow spanner indicates a task requiring some knowledge of diesel engines.

The purple spanner warns you that this task is more complex and requires advanced mechanical knowledge

Simple
● Time
● Tools

Technical
● Time
● Tools

Complex
● Time
● Tools

Notes about estimated time required; material and tools needed.

Checking the engine oil level

Simple

- 2 minutes
- Rag

Monitoring the oil level is the most basic of all precautions.

Important

Never sail with the oil at the 'minimum' level or below minimum. Such negligence makes the oil work much harder.

> An engine in good condition may consume between 0.2 and 0.5 litres of oil for every 20 hours of operation. It is important to top it up. Do not wait for the warning light to come on before checking the level!

Method

The engine oil level must be checked before the first start of the day (cold), or wait around five minutes after stopping the engine.

The oil level is checked with a dipstick which has two markings: one on top – the maximum level, one below – the minimum level. This dipstick is generally located on the engine block.

Procedure

Pull the dipstick out, wipe it off with a clean rag, then push it all the way back in. Pull the dipstick back out and check the level of the oil. Make sure the oil level is towards the maximum mark on the stick. The oil should never go below the minimum mark.

Caution: when checking the oil level, make sure the dipstick is fully inserted.

> **! IMPORTANT**
> Too much oil (level above maximum mark), is never desirable. It can cause over heating, loss of performance, leaks and an increase in temperature which is detrimental to good engine operation.

1 To check the oil level efficiently, wipe off the dipstick thoroughly then fully insert it again.

2 The level should be within the zone etched on the dipstick. The lowest mark is the minimum level, the highest, the maximum level.

3 If you have to add oil, don't forget to wait a few minutes before checking the level to allow time for the oil to get down to the sump.

4 When checking the level, also check the oil colour to discern any possible trace of water. The oil in this case becomes yellowish and contains small water bubbles. If your oil level is clearly above the maximum level, black in colour and more fluid, this could mean bad atomisation at the injectors; the injected diesel does not fully ignite and goes down to the sump via the piston rings.

Changing the engine oil; replacing the oil filter

Technical

- 1 hour 30 minutes
- Suction pump, oil filter wrench, basic tools, rags, detergent

The oil change is an important task in your engine maintenance programme. The engine's moving parts are subjected to friction, causing heat capable of fusing the metals in contact. The engine oil has the job of reducing this friction. Subjected to extreme thermal strains, the oil breaks down and, during the operating period, becomes charged with numerous metallic impurities as well as combustion residues which give it that blackish look. Therefore it should be changed periodically.

Frequency

Until about 1970, we changed engine oil every 50 hours. With improvements in oil quality, air and oil filters, manufacturers have changed their recommendations and as a general rule, on engines under 50 horsepower, they only recommend a change every 100 or 200 hours or every year. In any case, it is important to change the engine oil as recommended by the manufacturer. Caution: changing the oil is necessary but dirty, so prepare for this job carefully.

Preparations and precautions

It is necessary to know the quantity of oil to be used, so check your manual or contact the manufacturer. If changing the filter, allow an extra 0.2 to 0.5 litres.
   Gather the specific tools needed for this task:

- Pump or oil syringe
- Filter wrench
- Used oil container

Also prepare the parts and products to be changed:

- Oil
- Oil filter
- Other items needed (degreaser, rags, etc)

Locate all the points where you will be working:

- Oil filler cap, gearbox
- Fill openings
- Oil filter, etc

How to proceed

First, a rule to remember: an oil change must always be done while the engine is hot.
   The oil is removed using a pump inserted into the dipstick chamber or an opening designed for this purpose.

 Some engines have a built-in pump. A small hose fitted on it will avoid dirtying the bilge.

- Remove the dipstick.
- Connect the inlet hose of the drain pump to the opening.
- Open the filler cap to let the air in.
- Place the pump outflow hose in a receptacle (empty used oil can).
- Drain the oil 5 minutes after the engine has been turned off.
- Pump until the oil sump is dry.

Changing the oil filter

The oil filter's task is to catch the impurities the oil has accumulated from the engine's mechanical wear. It is therefore essential to change the filter at each oil change to keep these impurities, mostly consisting of metal particles, from travelling through the lubrication lines and acting like abrasives between the moving parts.

- Remove the oil filter with the filter wrench (a screwdriver can also be driven through the filter just off centre and used as a lever).
- Place a little bucket and rags under the filter to avoid a mess when unscrewing the filter.
- Wipe off and check the seat of the seal on the engine block.
- Lubricate the seat surface of the new filter seal.
- Hand-screw the new filter in, until it comes into contact with the engine block, then tighten one half to three quarters of a turn.
- Remove the dipstick to create an air intake and facilitate re-filling.
- Fill the oil sump up to the maximum level mark on the dipstick.
- Run the engine two to three minutes at idle.

 PRECAUTION
Check that the oil pressure warning light or alarm turns off after a few seconds and that there are no leaks around the oil filter.

- Stop the engine.
- Check the oil level two or three minutes after stopping.
- Top up as needed.
- Clean up the engine compartment and remove any traces of oil. Make an entry in the engine log book (date, type of oil, filter, number of hours).

> **! WARNING**
> The used engine oil is an environmental pollutant. Do not dispose of it carelessly. Once emptied into a tin or bucket, take it to the collection containers provided for this purpose in every marina.

Which brand of oil?

During the warranty period, you must use the recommended oil brand. After that, you may choose your own. However, be sure to maintain the same quality and viscosity, at least when changing oil.

Which oil quality?

Good quality oil is essential for good engine operation. It is wise to choose its 'quality' but also its viscosity in accordance with the manufacturer's recommendations. The API or CCMC classification serves as reference to assess the oil's intrinsic quality (see pages 44–5).

The SAE classification grades the oils according to their viscosity index. Nowadays all oils are multigrade. Their viscosity is suitable for winter as well as summer use. The choice of grade depends on the boat's usual operating environment. Today, most manufacturers recommend oils with a viscosity of 10 W 40 and having a minimum classification of CD (API Standard)

2 Remember to remove the filler cap to create an air intake.

3 Pump out all of the oil contained in the engine's oil sump. The oil change is done with the engine hot to make it easier to suck out the oil.

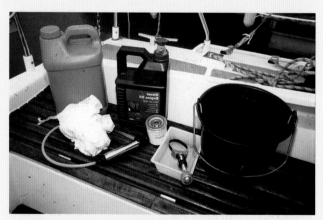

1 Before starting, gather the tools needed. Prepare the oil and the oil filter.

4 Use a filter wrench to unfasten the oil filter.

5 Prepare a receptacle, some rags or a plastic bag to collect the oil inside the filter.

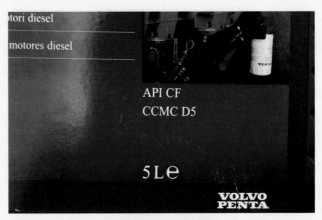

8 Carefully check the quality of oil indicated. Here the oil quality to be used is CF API standard or D5 CCMC standard – an excellent performance index.

6 Lubricate the rubber seal on the filter with a drop of new oil. Remember to clean the seat seal on the engine block.

9 When filling up, remember to remove the dipstick. This helps the oil flow down to the oil sump.

7 Never tighten the new filter with the filter wrench. Only hand tighten it until it is in contact with the engine block, then turn it an extra half to three-quarter turn.

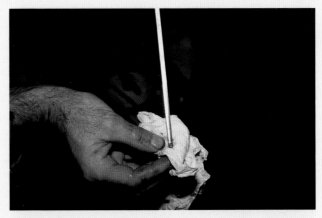

10 After running the engine for a short while, wait a few minutes before checking the oil level. Top up if necessary.

Checking the gearbox oil level

Simple

⬢ 5 minutes

⬢ Rag

This should be done on a cold engine, or wait for five minutes after turning off the engine.

The gearbox oil level is checked with a dipstick attached to a screw cap, which is screwed onto the gearbox case. Unscrew the cap and wipe the dipstick clean. Check the oil level by inserting the dipstick without screwing it in. The level is checked with the cap screw undone, simply resting on the opening.

Check the state of the seal on the cap. Screw the cap back on the casing. It may have one or two level marks depending on the type of gearbox.

Topping up or filling up is generally done through the dipstick opening.

 Only use oils recommended by the manufacturer. A different oil quality could result in extensive damage to the gearbox.

1 Unscrew the cap.

2 Wipe the dipstick clean, then re-insert it in the opening.

3 The level is checked with the cap unscrewed, resting on the opening.

4 Check that the level is between maximum and minimum. The oil should be translucent with no combustion residues. Take care: it is sometimes difficult to get a reading.

Changing the gearbox oil

Simple

- 30 minutes
- Oil pump or suction syringe, rag, funnel

This is done with the engine cold, either by unscrewing the drain plug, if easily accessible, or through the dipstick opening with the help of a suction pump (see 'Changing the engine oil').

Fill up through the dipstick opening with the type of oil recommended by the manufacturer. Check the level.

 On MS 2 Volvo boxes, the large filler cap at the top has to be removed to create an air intake.

1 To help drain the oil as well as when filling, remember to create an air intake. Here, unscrew the top screw cap. (Volvo gearbox)

2 Using the drain pump, draw out the oil contained in the box through the dipstick opening.

3 Follow the manufacturer's recommendations closely regarding the quality and quantity of oil for your gearbox.

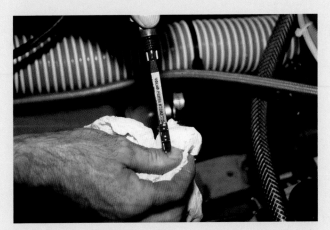

4 Wait a few minutes then check the oil level.

Changing the oil on an S-Drive transmission

Technical

45 minutes

Large flat screwdriver, funnel

When in 'dry dock', take the opportunity to change the oil. This can only be done with the boat out of the water.

Method

- Remove dipstick.
- Place a container under the transmission.
- Open the drain plug, located under the propeller's angle transmission casing.
- Let the oil drain.
- Check its colour. If water is mixed in the oil, it will have a grey colour and milky consistency, indicating a faulty seal at the gasket on the propeller shaft. The problem will have to be fixed before proceeding.
- Verify that the drain plug seal is in place and not damaged.
- Screw this plug back in when all the oil has drained.
- Tighten the plug.
- Re-fill with the manufacturer's recommended oil through the dipstick opening.
- Check the level (take care not to fill past the maximum level).

> On Volvo 120 S-Drive transmissions, the large cap on the top will have to be removed to aid filling by creating an air intake.

2 To unscrew the drain plug, use a screwdriver with a tip as wide as the slot on the plug. You may tap on the screwdriver with a hammer to ensure good engagement in the slot. Always use good quality tools for this operation to ensure an easy and accurate job. Here the screwdriver has a hexagonal section to allow use of a ring spanner to help free the plug.

1 To help draining and filling, remember to create an air intake.

3 Collect the oil in a container. Take your time and let the oil drain completely so all of it can be replaced.

4 The colour of the oil will tell you the state of water tightness in the lower leg. Also check for metal debris which could indicate abnormal wear of one of the gearbox components (pinion, ring, cone...). If any wear is present, consult an authorised dealer for your engine.

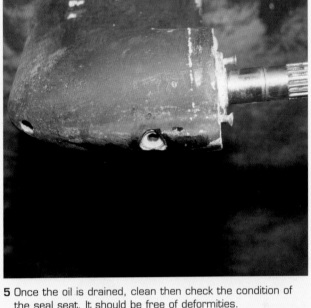

5 Once the oil is drained, clean then check the condition of the seal seat. It should be free of deformities.

6 Check the O-ring on the drain plug; replace it with a new one if you have any doubts or if there is the smallest crack. Also ensure that it is still supple and not perished.

7 The plug must be hand-screwed until it makes contact.

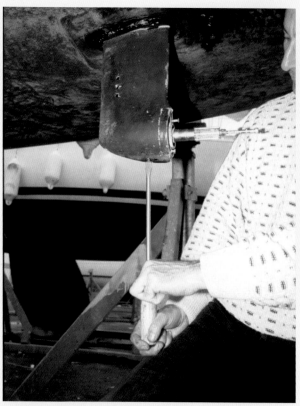

8 Tighten the plug. Some manufacturers recommend a specific tightening torque, eg Volvo: oil drain plug: 10 + or – 5 Nm.

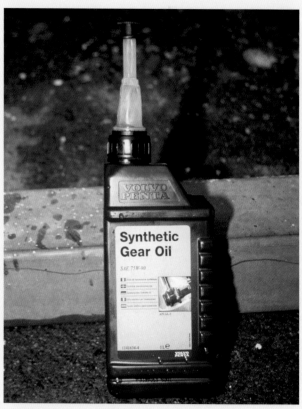

9 Always use the manufacturer's recommended oil. It will help preserve the lower leg and fulfil the manufacturer's warranty.

10 Fill the lower leg through the oil dipstick opening. A small funnel makes the job easy and keeps the bilge clean. For information concerning the quantity *and* quality of oil, consult the manual supplied by the manufacturer.

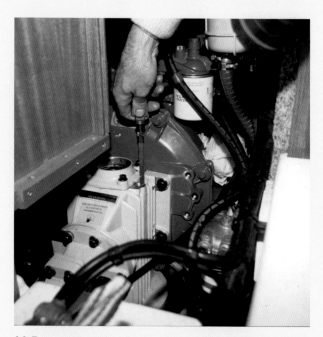

11 Do not screw the dipstick back on when checking the oil level.
    If the oil level exceeds the maximum mark, drain the excess oil.

Changing and cleaning fuel filters

Technical

- 1 hour
- Basic tools, rag, filter wrench

Diesel engines have one or more filters which require particular attention.

Frequency

The diesel engine's worst enemies are impurities and water contained in the fuel. For this reason, manufacturers put great emphasis on the cleanliness of filtering components. Many recommend changing the cartridges every 200 hours.

Check

It is wise to regularly check and, if necessary, bleed the primary water separating filter, located between the tank and the fuel pump. If the bowl is glass, it will be easy to see if there is water in the fuel. If there is water at the bottom of the bowl, you can remove it through the drain screw at the bottom of the filter.

> The filter may clog prematurely if poor quality fuel is used. This is easily identified by the loss of engine power. This power loss causes reduced boat speed as well as excessive exhaust smoke.

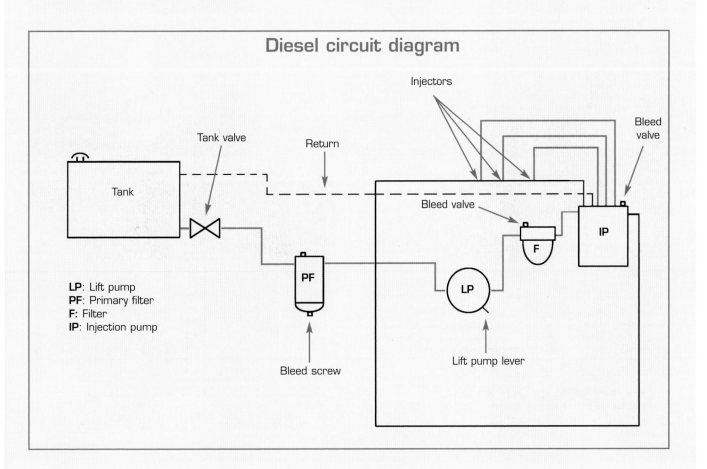

Diesel circuit diagram

LP: Lift pump
PF: Primary filter
F: Filter
IP: Injection pump

Changing the primary filter element

All jobs involving the fuel system require scrupulous cleanliness.
Be careful to avoid getting diesel on your hands.

◉ Method

- ◆ Close the fuel valve.
- ◆ Loosen the central screw.
- ◆ Pull out the bowl with caution, as it contains fuel.
- ◆ Remove the element and throw it away.
- ◆ Clean the bowl.
- ◆ Fit a new filter.
- ◆ Check that contact surfaces are clean.
- ◆ Replace the seals.
- ◆ Reassemble the filter, and moderately re-tighten the assembly screw.
- ◆ Open the fuel valve.
- ◆ Bleed the fuel line.

1 First, locate the primary filter. It is always fitted between the tank and the fuel pump. Its role is to filter out the coarser impurities. A screw at its base allows you to purge the water.

2 Unscrew the central screw on the filter. Prepare some rags and a container to collect the fuel.

3 Disassemble the bowl carefully as it contains fuel.

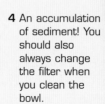

4 An accumulation of sediment! You should also always change the filter when you clean the bowl.

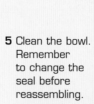

5 Clean the bowl. Remember to change the seal before reassembling.

6 Put a new filter in its cup. Check that the contact surfaces are scrupulously clean. Reassemble and tighten the filter.

Replacing the filter

◆ Close the fuel valve.
◆ Unscrew the filter with the help of a wrench designed for that purpose.
◆ Empty the filter and dispose of it.
◆ Check the seal seat.
◆ Screw on the new filter.
◆ Hand tighten it until the seal comes in contact with the cover, then tighten an extra half turn.
◆ Bleed the line.

1 Locate the filter. It is generally situated at the fuel line's highest point.

2 On a Yanmar engine, as shown in the photo, lightly tap on the filter nut with a hammer and chisel to loosen it. On other makes, loosen the cartridge with a filter wrench.

3 Remove the bowl by tilting it slightly.

4 Remove the filter. Have a small receptacle ready to collect the fuel and the filter.

5 Thoroughly clean the bowl.

6 Put in a new filter. Use only original parts as they ensure proper engine function and maintain the validity of the factory warranty.

7 Put the bowl with its seal in place, and screw on the nut.

8 Lghtly tap on the filter nut with a hammer and chisel to tighten it. With other makes, you may be required to hand tighten the cartridge until it comes in contact with its seat, then give it an extra half to three quarter turn.

Cleaning the fuel pump gauze

The fuel pump on the engine may be fitted with a filter gauze under its cover. This will have to be cleaned at least once every season.

- ◆ Note the filter gauze's position before removing it.
- ◆ Unscrew the central screw and remove the gauze.
- ◆ Clean, rinse and dry the filter gauze.
- ◆ Clean any deposits in the fuel pump body and cover.
- ◆ Replace the gauze in its original position.
- ◆ Tighten the central screw.
- ◆ Bleed the line.
- ◆ Immediately after starting the engine, check to make sure that there are no fuel leaks. If there are any leaks, air will enter the system and cause problems during operation.

Some types of pump, especially those fitted on Volvo engines, have a gauze filter that must be removed and cleaned.

Bleeding the fuel system

Technical

- 30 minutes
- Basic tools

The fuel system must be bled:

◆ Whenever work done could cause an air leak in the system, for example, changing a filter.
◆ When the tank has been completely emptied (running out of fuel).
◆ When starting the engine after a long idle period or when starting for the first time.

The way the lines are bled may vary depending on the engine type. Some injection systems are self bleeding, provided the starter is allowed to crank long enough.

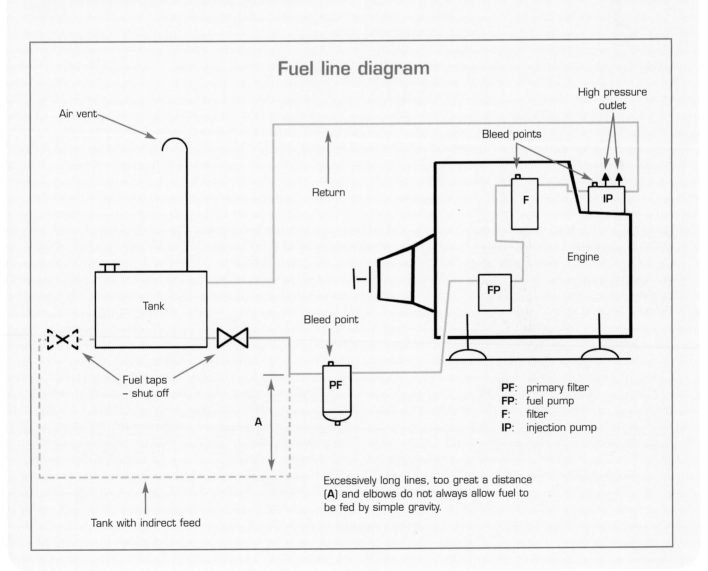

Fuel line diagram

High pressure outlet

Air vent

Bleed points

Return

F

IP

Engine

Tank

FP

Bleed point

Fuel taps – shut off

PF

A

PF: primary filter
FP: fuel pump
F: filter
IP: injection pump

Tank with indirect feed

Excessively long lines, too great a distance (**A**) and elbows do not always allow fuel to be fed by simple gravity.

Bleeding the primary filter

- Open the fuel tap.
- Wrap a rag around the filter, then open the upper bleed screw or unscrew the outlet connection two or three turns.
- As soon as the fuel runs clear, without bubbles, close the bleed screw or tighten the outlet fittings. Wipe the filter dry.

The primary filter can also be primed by hand-operating the fuel pump. Proceed as follows:

- Open the fuel tap.
- Unscrew the fuel pump outlet fitting two or three turns. The bleed screw and the outlet fitting on the primary filter must not be open.
- Operate the fuel pump to draw the fuel through. This could take a few minutes.

If you aren't able to bleed the primary filter, there may be several reasons:

- The fuel tap is closed.
- The tank is almost empty.
- The tank fuel line is indirect.
- The tank fuel line is empty.

To facilitate bleeding the line:

- Locate the air vent on the tank.
- Disconnect the hose.
- Blow and maintain pressure (tiring). Blow, then fold hose to maintain pressure (easier).
- Repeat the operation until the line is primed.

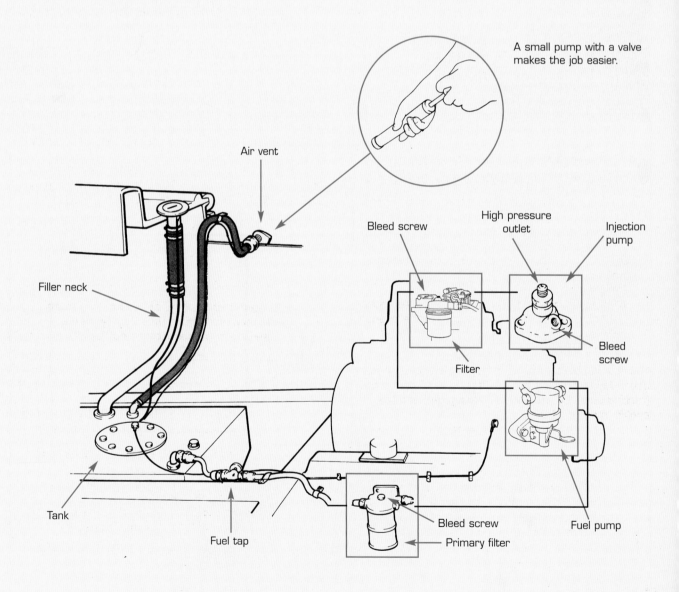

A small pump with a valve makes the job easier.

Air vent

Bleed screw

High pressure outlet

Injection pump

Filler neck

Bleed screw

Filter

Tank

Fuel tap

Bleed screw

Primary filter

Fuel pump

Bleeding the filter

◆ Unscrew the bleed screw located on top of the filter one or two turns.
◆ Operate the lever on the fuel pump until the fuel flows free of water.
◆ Close the bleed screw.

Open the bleed screw

Bleeding the injection pump

◆ Unscrew the bleed screw located on the top of the injection pump one or two turns.
◆ If the filter does not have a bleed screw, unscrew the outlet fitting.
◆ Operate the fuel pump lever until the fuel flows free of water.
◆ Retighten the bleed valve.
◆ All injection pumps have at least one bleed screw. Some have two. The upper one must be opened first, since air naturally escapes through the highest opening.

You should now be able to restart the engine. It is a good idea to position the throttle on maximum revs (making sure the propeller is disengaged and the engine 'stop' is all the way in).

If the engine does not start, then the high pressure line must be bled.

 Sometimes the fuel pump lever doesn't work. The solution is to let the engine turn over. Half a turn is usually enough to free the cam blocking the lever.

Bleeding the high pressure line

◆ Open the connector(s) on the pipes going to the injectors.
◆ Position the throttle lever on maximum.

Bleeding the high pressure line

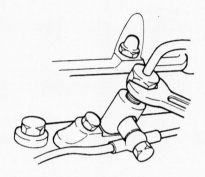

To bleed the high pressure line, open the connector.

◆ Precaution

◆ Place some rags under the injector connections to catch the fuel.
◆ Shut the seacock to avoid flooding the exhaust.
◆ Hit the starter a few times or hand crank it while decompressing, until fuel flows out without water or air bubbles.
◆ Bring the throttle back to 'fast idle'.
◆ Retighten the fitting(s).
◆ The engine is now ready to start.

 On an in-line pump, all connectors must be loosened, while on a rotary pump, opening one connector is sufficient.

! IMPORTANT

Before laying up the boat for the winter, there are four basic recommendations:

◆ Fill up the tank to avoid condensation inside the tank;
◆ Bleed the water separator filter;
◆ Replace all fuel filters;
◆ Bleed the fuel line, let the engine run for a while and look for any leaks.

Changing the anodes

Simple

⬡ 30 minutes

⬡ Basic tools

To protect the engine against galvanic corrosion, it is important to install zinc anodes at critical spots on the engine: the cylinder head, heat exchanger and engine block.

⬥ Frequency

As a general rule, the anodes must be replaced every 200 or 300 hours or yearly. In the interim and during inspections they can be carefully brushed with a wire brush. If they are more than 50% worn, they must be replaced.

◆ The first precaution to take is to turn off the main switch or disconnect the battery. Shut the seacock and drain the cooling system.

◆ Remove the anode from its support and replace with a new one.

◆ Open the seacock.

◆ Check the anode's seal when starting and once the engine is hot.

2 Unscrew the old anode from its support. Take care to keep its seal (Yanmar).

3 Compare this used anode to a new one. Not much left! It is almost totally corroded.

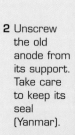

1 Locate, then unscrew the anode.

4 Once the anode has been replaced, screw and tighten it back on the engine block.

Checking and replacing the drive belt

Technical

 30 minutes

Spanner, screwdriver, belt

Although they last much longer nowadays, a belt breaking is still one of the main causes of breakdown.

This is a breakdown that is usually due to lack of maintenance. The belts fitted on our engines are made of a mixture of natural and synthetic rubber. They are cast around a fabric of synthetic fibres.

There are two types of belts:

◆ The trapezoidal belt is the type most commonly used to drive the alternator or the engine cooling water pump.
◆ Much flatter, the notched belt also called Poly V, carries the load on both sides and has good flexibility. For this reason, it is often used on larger engines to drive multiple devices.

A belt's life span depends on the cleanliness of the engine bay, but most of all, on its correct tensioning.

A belt operating in water mixed with oil wears very quickly.

Too little tension and it risks slipping on the pulleys, which results in premature mechanical wear and excessively high temperatures. The function and efficiency of the alternator and cooling system are greatly affected by slipping belts. Too much tension and the belt can be damaged as well as the alternator or water pump bearing. It is therefore important to check belt tension and wear several times a year.

Checking the belt

First, check that the belt has no cracks by twisting it. If it is cracked, a new belt must be refitted. Then check its tension by pushing on the side of the belt between the pulleys that are farthest apart. When pushing on the belt between the pulleys with the thumb, it should not give more than an amount equal to its thickness (5 to 10mm).

Adjusting the belt

The adjustment is made by pivoting the alternator on its fastening brackets. To do this, proceed as follows:

◆ Isolate the power supply.
◆ Loosen the fastening bolts on the alternator or on the seawater water pump (Yanmar GM).
◆ Push the alternator back with a lever.
◆ Check the belt tension by pushing on it. The amount of give must not be more than its thickness (5 to 10mm).

◆ Retighten the fastening bolts.
◆ Turn the main switch back on.

 A slack belt emits a characteristic squeal. If this is the case, the belt must be re-tensioned without delay because it won't do its job properly. The belt will slip, heat up and eventually break.

A belt that is too tight causes excessive wear on the alternator and water pump bearings. If the belt is cracked or appears brittle, it will have to be replaced without delay.

The belt must always be put on with the tensioning bracket in its minimum position, so as not to force the pulleys or the belt.

After a few hours of engine running, check the tension again and if necessary, re-adjust.

Fitting a new belt

If the belt breaks, the charge warning light should come on immediately. Stop the engine as soon as possible. When the belt that drives the water pump is broken, the engine cooling system no longer functions.

Always carry a spare belt on board; it should be part of the engine's safety equipment.

Method

◆ Make sure the power supply is off.
◆ Unscrew the alternator fastening bolts.
◆ Push the alternator in as close as possible to the engine. When putting on a belt, the tensioning bracket should always be in its minimum position to avoid forcing the pulleys and the belt.
◆ Put on the new belt, using a screwdriver as a lever if needed, but taking care not to damage the belt.

If the belt won't go on by hand, check that it is the correct size and that the mounting bracket is loose.

- Push the alternator away from the engine with a lever.
- Check the belt tension.
- Retighten the fastening bolts.
- Turn the power supply back on.

1 Check the belt's tension as well as the general condition. It is important to change the belt at the first sign of a crack.

2 Be sure to also check the sides of the belt for wear. An under-tensioned belt slips and will soon wear.

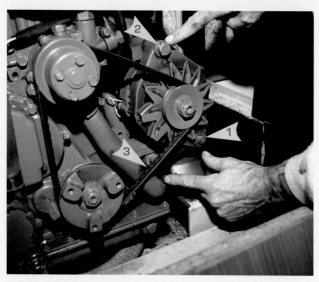

3 Locate the alternator fastening bolts (1, 2 and 3). Here, the three points have to be loosened to be able to pivot the alternator.

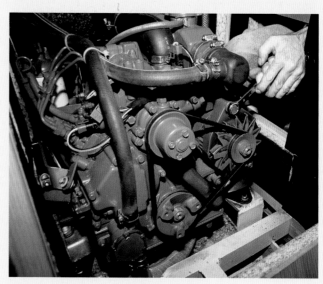

4 Loosen bolts 1, 2, and 3 (see photo 3).

5 Push the alternator away from the engine with a lever. Tighten bolt **1** first. Then tighten bolt **2** and **3**.

6 To remove the belt, the alternator has to be pushed in completely to free the belt.

7 Remember to clean the pulley grooves before putting a new belt on.

8 Tension the belt.

9 Check the tension after running for a few hours.

Servicing the direct cooling system

Simple

- 1 hour
- Basic tools

With time, scale, rust and salt build up inside the cooling system. The result is a reduction of the water passages, hence the engine tends to overheat. Therefore, the cooling system components have to be inspected regularly, eg: water pump, thermostat, etc.

Draining the seawater line

- Close the seacock.
- Open the drain plug (beware, sometimes a build-up of scale or salt keeps the water from flowing out. This obstruction of the drain opening will have to be removed).
- Screw the drain plug back on.

Flushing, protecting and winterising the seawater cooling system

To keep the cooling system free of accumulated deposits and salt crystals, it must be flushed with fresh water. This simple operation can be done on the water or during winter storage on the hard.

Method

- Flush the engine with fresh water after having disconnected the inlet hose. Submerge the inlet hose in a bucket filled with fresh water.
- Run the engine for 30 minutes.
  There are two options:
  - you either decide to drain the pipeline, including the muffler;
  - or, to limit internal oxidation, you fill a bucket with a four-season coolant. You then submerge the inlet hose in the bucket with the engine running. Filled with four-season coolant, which limits oxidation, the pipeline and muffler no longer need to be drained.
    Re-connect the hose to the seawater valve.

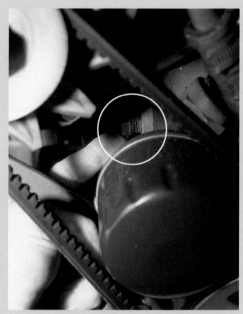

1 It is sometimes difficult to find, and access, the drain plug on the engine block. It is generally situated on the lower part of the engine block below the water passages.

2 Drain the cooling system after turning the engine off.

3 Disconnect the hose on the seawater pump and submerge it in a bucket that you must continuously replenish with fresh water. Let the engine run for a few minutes. Adjust the flow rate of fresh water with that of the engine's water consumption and keep an eye on the level in the bucket. Never run it dry as this would cause extensive damage to the water pump impeller.

4 Fill the bucket with a four-season coolant.

5 When the four-season coolant begins to flow out through the exhaust, stop the engine. Your engine is now protected.

Servicing the indirect cooling system

Technical

- 1 hour
- Basic tools

In an indirect cooling system, the four-season coolant circulates through the heat exchanger where it is cooled by seawater. The cooling temperature is regulated by a thermostat on the cylinder head, the hottest part of the engine.

Flushing the fresh water circuit

To maintain the cooling system in perfect working order and avoid any loss in cooling performance, it is important to keep it free of deposits and impurities by regularly flushing the fresh water circuit.

Method

- Before doing anything else, disconnect the battery or turn the engine contact off.
- Then, drain the line with the drain plug or valve situated on the engine block. Remove the cap on the heat exchanger to create an air intake.
- Screw the drain plug in or close the seacock.
- Fill the system with fresh water, adding a good, non-foaming detergent.
- Run the engine for at least 15 minutes then drain again.
- Fill the system with fresh water and run the engine for 5 minutes. Drain.
- Repeat this operation until the drained water is clear and clean.
- Empty the last lot of water then re-fill again with an antifreeze-water mix or a four-season radiator coolant.
- Let the engine run for a few minutes, accelerating now and then.
- Turn off the engine and top up again.

> ✸ **Some manufacturers recommend that you do not remove the heat exchanger cap while the engine is hot and the temperature in the system is still high. In this case, the service should be done when the engine is cold.**

If there are excessive deposits or persistent impurities, the heat exchanger core will need to be disassembled. Once disassembled, the best solution is to submerge the core in a manufacturer-approved non-caustic cleaning solution.

When the tubes are clogged by soft deposits or debris, insert a steel rod in the tubes in the opposite direction of the water flow. Take care not to damage the tubes.

1 Locate the drain plug on the exchanger or the engine block.

2 Unscrew the plug, place a container under the drain hole to catch the flow and to avoid a mess.

3 Drain. Note the rusty colour of the liquid and the slow flow due to partial obstruction of the drain hole by deposits.

4 Prepare the detergent cleaning solution.

5 Fill the container with fresh water. Shake vigorously to thoroughly dissolve the crystals

6 Fill the fresh water system, then run the engine for a minimum of 5 minutes.

7 Drain it again. Once more, note the colour and flow rate. The drain hole is unplugged, the cleaning solution is full of scale deposits and rust residues. Put the drain plug back in, then fill the freshwater system again. Run the engine for 5 minutes, drain again. Repeat this flushing operation until the drained water is clear and clean.

8 Drain one last time and fill with a four-season coolant. If your engine has a bleed screw, unscrew it, fill the system up then screw it back in. Then fill the exchanger to its usual level.

9 Clean the engine bay. Remember to check the fluid level in the heat exchanger after running the engine.

Checking the seawater pump impeller

Simple

- 1 hour
- Basic tools

The engine's water pump has a neoprene or rubber impeller. This has to be checked every 200 hours or at least once a year.
Proceed as follows:

- Close seacock.
- Remove the pump cover.
- Then remove the impeller.

 On some brands (older Volvo models), the impeller shaft has to be pulled out 10 to 15mm to remove the set screw that goes through the impeller.

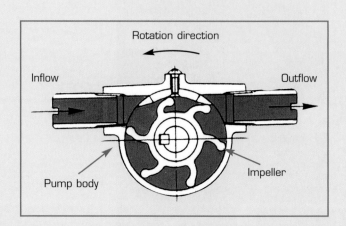

Exploded view of a seawater pump

Never hesitate to change components that don't seem to be in perfect condition. Your engine's longevity depends on it.

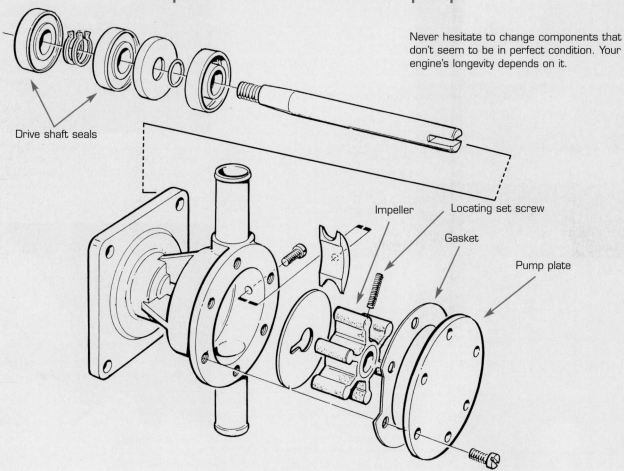

Drive shaft seals

Impeller

Locating set screw

Gasket

Pump plate

In a small engine compartment, it is sometimes easier to remove the entire pump, by disconnecting the hoses and fastening bolts.

◆ Check the state of the impeller. If it is even slightly damaged (torn, scratched, worn), do not hesitate to change it.

◆ When changing the impeller, remember its direction of rotation, ie the orientation of the impeller blades, when reassembling it.

◆ Grease the impeller blades and the pump body with glycerine.

> **! IMPORTANT**
> **In an indirect cooling system, if an impeller blade has been torn off, you will usually find it at the heat exchanger inlet.**

◆ Put the cover back on with a new gasket, after having carefully cleaned the seats (casing and cover).

◆ Open the seacock.

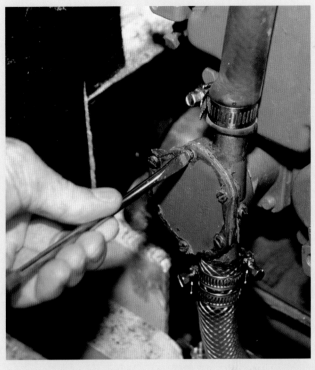

2 Tap on the screwdriver with a mallet to ensure a good grip in the slot. Unfasten the water pump cover plate screws.

1 On most engines, the water pump is easily accessible.

3 Tap laterally with the handle of the screwdriver or a mallet to loosen the cover plate. Never insert the screwdriver in the seal seats.

4 When the plate is removed, the impeller is exposed in its pump casing. Note the deep scratches on the plate. This one is well worn. To get yourself out of trouble, it is possible to smooth it with sandpaper on a flat surface or sometimes to turn it over, first taking care to remove any paint.

5 Extract the impeller with the help of two round-stem screwdrivers to avoid damaging the pump casing.

6 Inspect the impeller. Here, there is no doubt: the blades are torn off. The impeller has to be changed.

7 Here the damage is less obvious but we can still discern some signs of tearing at the base of the blades. To be safe, replace the impeller and keep it as a spare. Costly as it might be, never hesitate to replace components such as the cover plate, impeller and sometimes the pump casing, when they appear not be be in perfect condition. The longevity of your engine depends on it.

8 Most manufacturers recommend kits that contain the gasket and the impeller. Never use adaptable impellers such as the one shown in the photo except when you are unable to find the original part or in an emergency.

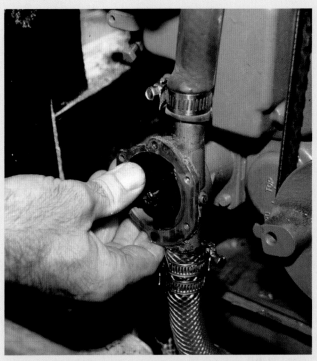

9 Wet the impeller with soapy water then replace it in the pump casing, checking its direction of rotation. Make sure the blades are properly oriented.

10 Put the cover plate with a new gasket back on the pump case.

11 Open the seacock and check that the cooling system works well. Ensure that there is no water seepage around the cover plate.

Checking the thermostat

Technical

- 1 hour
- Basic tools

The thermostat regulates the engine's operating temperature.

In a direct cooling system, the temperature must not exceed 55°. In an indirect system, the temperature should stay close to 90°. Therefore, it is important to check the temperature at which the thermostat opens every 500 hours of engine operation. To do this, you submerge it in a container filled with water and check its opening temperature with the help of a thermometer. If the engine's water temperature goes up abnormally (red zone occasionally reached, then back to its usual level), or if white smoke is emitted from the exhaust long after starting the engine, check the thermostat.

Method

Generally located on the upper part of the engine, the thermostat is easily removed.

- Close the water tap
- Drain some of the cooling water.
- Remove the thermostat housing.
- Remove the thermostat.

Testing the thermostat

- Place the thermostat in a container full of water after having noted its opening temperature (imprinted on it).
- Heat the water.
- Check to see at what temperature the thermostat opens. If the result is different, it will need to be replaced.
- Clean the seat on the housing.
- Reassemble the thermostat with a new gasket.
- Tighten the housing.
- If it is an indirect cooling system, top up the system.
- If it is a direct cooling system, open the seacock.
- Ensure that the cooling water circuit works properly and is watertight.

1 To access the thermostat, unfasten the hose clamp and unscrew the bolts on the thermostat housing located on the cylinder head.

2 Gently loosen the housing with the help of a mallet and remove the thermostat. Check the inside of the hose. If the rubber is perished or excessively scaled, replace it.

3 Clean the thermostat and read its opening temperature. Here, it is 82°. Examine the thermostat and look for any deterioration. If it is damaged, open when cold, or is excessively scaled, it must be replaced.

4 Submerge the thermostat in a container filled with water. Here, a baby-bottle warmer does the job. Observe the opening temperature and the valve travel.

5 If it doesn't open, or if the opening temperature and valve travel do not meet the manufacturer's specifications, replace the thermostat.

6 Verify that the valve sits well in its seat in its resting/closed position.

7 Ensure that the seal seats are perfectly clean, and put the thermostat back in place.

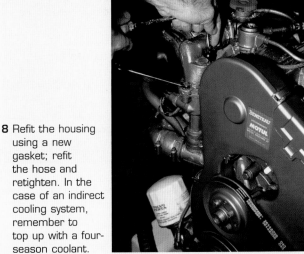

8 Refit the housing using a new gasket; refit the hose and retighten. In the case of an indirect cooling system, remember to top up with a four-season coolant.

Checking the heat exchanger cap

Simple

- 5 minutes
- Screwdriver, cap pressure gauge

If your engine consumes cooling fluid and its operating temperature is too high, the first thing to check is the heat exchanger cap.

When the heat exchanger cap leaks, the cooling system does not operate under pressure. The four-season coolant then boils at a lower temperature than the manufacturer's specification and escapes in great quantities.

Method

- Remove the heat exchanger cap.
- Check its seal. It must show no trace of damage or wear.

- With the help of a small screwdriver, delicately lift the valve at the centre of the cap. This valve must sit perfectly in its seat with light pressure.
- Change the cap if necessary.

 The maximum allowable pressure is engraved on the cap.

- Check that this corresponds to the manufacturer's specification. If in doubt, consult a specialist to check the calibration of the spring with a special tool.

1 If your engine is hot, open the cooling system cap with caution (wearing gloves and goggles) as hot steam or four-season coolant could escape – be very, very careful. Turn the pressurised cap to its first notch to let the pressure off. Remove the cap.

2 The correct setting for the pressure relief valve is engraved on the cap. The setting indicates the pressure at which the valve will open to prevent the engine from boiling over. This improves combustion and engine performance.

3 Check the state of the seal.

4 Also check that the cap return valve is in good working order by carefully lifting it with a screwdriver. The return valve prevents a vacuum from forming in the cooling system when the engine is shut off and begins to cool.

5 A cap pressure gauge allows you to artificially put the cooling system under pressure. It is also very useful for detecting small leaks.

6 It also allows you to check the pressure at which the pressure relief valve opens.

Checking the battery

Simple

- 15–20 minutes
- Battery hydrometer, multimeter

A battery is an electrical reservoir. It is designed to supply electricity to the various components and accessories needed for the boat's normal operation. It is usually recharged by a generator: the alternator.

However, during heavy energy consumption or during winter storage, the battery's potential decreases. The battery's charge state must be checked every 100 hours or at least once every season.

● Checking the electrolyte level

On traditional batteries, the electrolyte level has to be checked regularly. The high current at the end of charge causes water to decompose into gases (hydrogen and oxygen) that escapes through the vents on the battery caps. It must therefore be topped up with demineralised or distilled water.

The electrolyte level must be around 1cm above the plates. The majority of batteries have a window so you can see this. Excessive water consumption could be a sign of an over-heating battery due to regulator maladjustment.

> **! IMPORTANT**
> When adding water during winter storage, charge the battery immediately to avoid the risk of freezing where water and acid may not be mixed properly. Some so-called 'maintenance free' batteries need a layer of water above the ice film to prevent them from completely drying out.

> **! WARNING**
> To avoid any risk of explosion, never check the fluid levels when holding a lighter or a match. Use a battery-powered torch (flashlight). To avoid producing sparks, the negative terminal should always be connected last and disconnected first.

1 Checking the levels of the electrolyte is a tiresome but necessary job as the electrolyte evaporates during the battery's charge and discharge cycles.

2 If the level of electrolyte is low, add only distilled or demineralised water. Use a small funnel for this job to avoid spilling water everywhere.

3 Ensure that each battery element is covered by about 1 to 1.5cm of water

Checking the battery charge state

The battery's state of charge must be checked at least once per season. The best way of doing this is to measure the density of the electrolyte with the aid of a battery hydrometer.

● Checking with a battery hydrometer

The hydrometer tells you the electrolyte density. The density varies according to the battery's state of charge.

● Method

◆ Check each cell individually. Use the pipette to draw up just enough water for the float to rise freely. Take care not to let any of it drip on the floor or on to your clothes as the acid is extremely corrosive.

◆ Hold the hydrometer vertical and at eye level.
◆ Read the density.
◆ A density of 1.28 indicates the battery is fully charged.
◆ A density of 1.20 tells you the battery is charged to only 50% of its capacity and needs to be charged.
◆ At 1.1 density, the battery has only 20% of its capacity and needs recharging urgently.

Most hydrometers only have coloured zones on the float to indicate the battery's charge state. For a more accurate reading, professionals use a compensation thermometer which indicates the correction to apply if the battery temperature is above or below 15°C.

> **! IMPORTANT**
> **Never take density readings immediately after adding water to the battery.**
>   **Wear goggles when dealing with battery acid.**

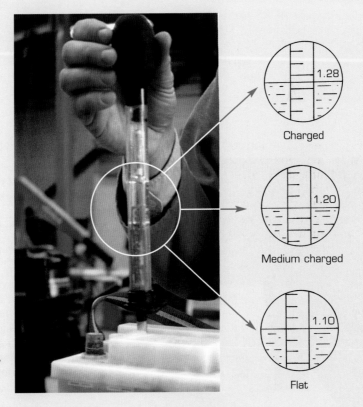

A battery hydrometer is an inexpensive tool that allows you to evaluate the battery's state of charge. Take a density reading for each cell. The weaker the density, the less the battery is charged.

1.28
Charged

1.20
Medium charged

1.10
Flat

● Checking with a voltmeter

If the electrolyte is not accessible (maintenance-free battery), a voltmeter graduated in hundredths of a volt must be used. This test has to be carried out when the battery is 'cold', ie when it has been at rest for at least two hours.

A reading of 13 volts indicates the battery is fully charged. At 12.2 volts, it is 50% charged and at 12.0 volts, it is charged at 25% of its capacity.

Some batteries (Freedom, Vetus), have charge indicators that allow the charge state to be checked at a glance:

◆ Green means the battery is charged.
◆ Black means the battery must be charged until the indicator turns green.
◆ Yellow or clear indicates there is a problem some-where (electrolyte too low for example).

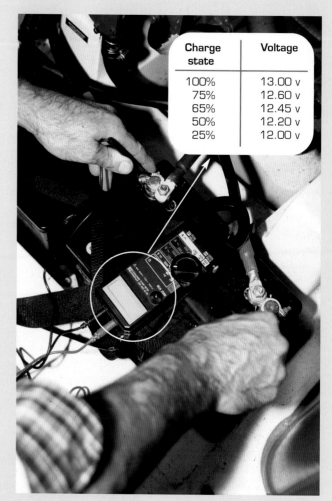

Charge state	Voltage
100%	13.00 v
75%	12.60 v
65%	12.45 v
50%	12.20 v
25%	12.00 v

2 Use a digital multimeter that measures one tenth or even one hundredth of a volt.

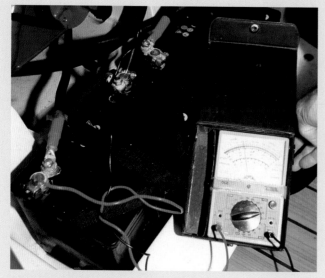

1 Reading the battery voltage is difficult and inaccurate if you use an analogue multimeter.

> **! IMPORTANT**
> ◆ **The battery terminals must not be reversed.**
> ◆ **Each terminal has a + or – sign. The negative terminal is earthed to the engine.**
> ◆ **To avoid any current leakage, you should always keep the top of the battery and its tray dry and clean. The terminals must be kept in good condition, without traces of sulphate or oxidation and should be well greased (Vaseline) and well tightened.**
> ◆ **Take care not to place metal tools across the two terminals.**

Recharging the battery

Simple

- 10 minutes to set up
- Battery charger

Apart from checking the electrolyte density or reading the voltage, you can tell if the battery is undercharged when the starter turns with difficulty or not at all when you try to start the engine. Charging is done with an electric charger.

- Remove the battery cell caps and check the fluid levels.
- Connect the + terminal (red wire) of the charger to the + terminal on the battery and the – terminal (black wire) of the charger to the – terminal of the battery.
- Before connecting the charger to the power supply, check that the voltage corresponds to the voltage of your battery. The old 6 volt batteries of the past have been superseded by 12 volt batteries.
- Set the charge level. This depends on the battery capacity. **Set it to two amperes per ten amperes of capacity**; no higher or there is a risk of battery deterioration.
- The end of the charging period is recognised by the bubbling of the liquid in all of the cells.
- Unplug the charger from the power supply.
- Disconnect the charger from the battery.

> **! IMPORTANT**
> **To avoid the risk of an explosion, never disconnect a battery without first disconnecting the charger from the electrical supply.**

Caution

The charge performance of a re-charged battery is around 75%: a 40Ah battery will take around 52Ah to charge. The maximum charging rate should not exceed 20% of the battery's nominal charge. If at the end of a long period of charge, the electrolyte density in all the elements does not exceed at least 1.20, the battery is not in good condition. If one of its cells will not recharge – density still below 1.1 on that cell – change the battery.

The expected life span of a battery is about four years. Old batteries gradually hold less and less charge.

1 Clean the battery terminals.

2 A little neutral grease (Vaseline) applied to the terminals inhibits the formation of salts.

3 Remember to remove the cell caps when charging the battery.

4 Do not exceed the charge density. Example: a 45Ah battery = maximum admissible 9A.

Adjusting the valve clearances

Technical

1 hour

Basic tools, feeler gauge

To get the best performance from your engine, high quality fuel and well adjusted injection are not enough. The valve clearances also have to be correct, ie meet the manufacturer's specification. This should be checked at intervals usually every 200 hours (see your engine manual) and any time you hear excessive noise coming from the cylinder head when the engine is running.

> ✹ **Engines fitted with hydraulic pushrods require no adjustment at all.**

The clearances, designed to let the valve stems move freely during operation, are given in the manufacturer's specifications for the intake and exhaust valves when cold (that is, at least six hours after the engine has been turned off). The clearance is often greater for the exhaust valves than for the intake valves. This is because of the different expansion of these parts when working at different temperatures.

The valve clearances are checked and set differently, depending on the engine type and make. This job does not require specific tools other than a feeler gauge, screwdriver and a suitable spanner.

Too large a clearance causes incomplete charging of the cylinders, affecting engine performance (the valve opens too late and closes too soon). Conversely, too small a clearance can result in compression loss and rapid deterioration of the valve heads and their seats.

On some modern engines – eg overhead cam – the manufacturer no longer requires this adjustment except to solve certain technical problems.

● Procedure

The valve clearance is generally set with a headless adjusting screw and lock nut, situated at one end of the rocker arm. The clearance is measured by sliding the blade of a feeler gauge in between the valve stem and the rocker. When an overhead cam acts directly on the valves, the feeler gauge is inserted between the lifter and the camshaft. If the clearance is correct, the gauge should just barely slide. Slight resistance must be felt when pulling it. Otherwise, an adjustment is needed.

1 Unfasten the screws and remove the rocker cover, disconnecting any breather hoses connected to it.

2 If it is stuck, free it by tapping it with a wooden mallet. Never insert a screwdriver between the seal seats.

3 Before checking and adjusting the valve clearances, identify the intake valves and the exhaust valves. Select the gauge corresponding to the desired clearance.

How to make the adjustment

◆ To access the valves, remove the rocker cover with care. It is generally bolted down at its centre with two or three bolts or screws.
◆ Identify the valves. To do this, simply see which valve faces the intake or exhaust duct.
◆ All adjustment methods have an essential prerequisite: the valve to be adjusted must be in its closed position.

Basic adjustment method

◆ Turn the engine over so that the valve to be adjusted is in its full open position.
◆ Turn the crankshaft over one full turn or the camshaft a half turn; the valve is now fully closed or on the crest of the cam.
◆ Adjust this valve.

Because of the need for numerous rotations for multi-cylinder engines, this can become a very tedious task unless it is tackled systematically – see page 122.
Manufacturers indicate in their manuals the method to use according to the type of engine in question. The general principles are explained below:

Setting the valve clearances

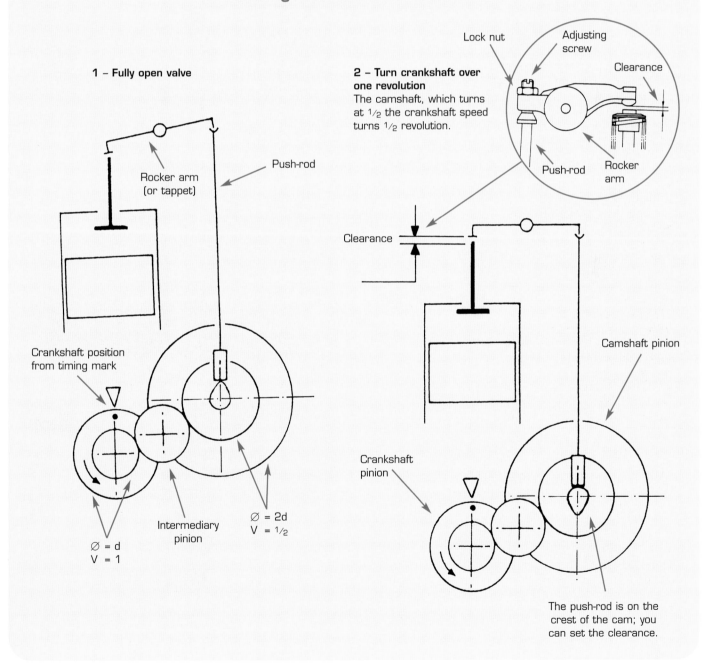

1 – Fully open valve

Push-rod

Rocker arm (or tappet)

Crankshaft position from timing mark

Intermediary pinion

$\varnothing$ = d
V = 1

$\varnothing$ = 2d
V = $^1/_2$

2 – Turn crankshaft over one revolution
The camshaft, which turns at $^1/_2$ the crankshaft speed turns $^1/_2$ revolution.

Lock nut

Adjusting screw

Clearance

Push-rod

Rocker arm

Clearance

Crankshaft pinion

Camshaft pinion

The push-rod is on the crest of the cam; you can set the clearance.

Adjustment method for 1, 2 and 3 cylinder engines

◆ Crank the engine over by hand, until the valves of the cylinder in question are in balance (end of exhaust, beginning of intake). Both valves are now slightly open. The piston in that cylinder is at TDC (top dead centre).
◆ Give the crankshaft one turn.
◆ Set the clearances for the valves on that cylinder.
◆ Perform the same operation for the other cylinders, bearing in mind the engine's firing sequence.

For a 3-cylinder engine, this method requires a considerable number of engine rotations. For that reason, some manufacturers such as Yanmar or Volvo use a two phase method of adjustment that is much faster.

◆ *Phase 1*: bring cylinder 1 to TDC of the compression stroke, and set the clearances on the intake and exhaust valves of that cylinder, as well as the exhaust valve of cylinder 2.
◆ *Phase 2*: turn the engine 240°, following its direction of rotation. Set the clearance of the intake valve on cylinder 2 and the intake and exhaust on cylinder 3.

Adjustment method for setting the valve clearances on a 4-cylinder engine

Principle

The most common method requires finding the cylinder whose valves are overlapping, also known as 'rocking', ie the piston is at the end of exhaust and beginning of intake phase. At this same time, another cylinder is at the end of compression-injection and beginning of explosion phase. These two valves are closed and free. We can then measure the clearance and adjust it if needed.

Adjustment method

◆ Crank the engine by hand until the valves on cylinder 1 are rocking (cylinder 1 towards flywheel end).
◆ Adjust the valve clearances on cylinder 4.
◆ Turn the crankshaft a half turn; cylinder 3 is then rocking.
◆ Adjust the valve clearances on cylinder 2.
◆ Turn the crankshaft a half turn; cylinder 4 is rocking.
◆ Adjust the valves on cylinder 1.
◆ Turn the crankshaft a half turn; cylinder 2 is rocking; Adjust the valves on cylinder 3.

Reassembly

◆ Put the rocker cover back on using a new gasket, after having carefully cleaned out all traces of old gasket on the cylinder head as well as on the cover.
◆ Tighten the fastening bolts a bit at a time. Caution: excessive tightening can cause irreversible damage to the cover and create oil leaks.
◆ Start the engine and make sure no oil leaks appear around the rocker cover.

A different method can be used while following the same cylinder order.

By putting the exhaust valve on each cylinder in full open position, we can see that if cylinder 1 is in the exhaust phase, cylinder 4 is in the compression phase. The intake valve on cylinder 3 is closed. Cylinder 3 is in the explosion phase and exhaust valve 4 is closed. We can then adjust intake valve 3 and exhaust valve 4.

The table on pages 124–5 shows the steps to take.

Setting the valve clearances without a feeler gauge.

In some cases, due to lack of access, it is very difficult to check the clearances using a feeler gauge. Then proceed as follows:

◆ Loosen the adjusting screw lock nut.
◆ Tighten the screw until the rocker arm and the valve stem touch.

To set the clearance to the manufacturer's specification, you need to know the thread pitch on the adjusting screw in order to determine the number of degrees to loosen it.

The following is an example of how to calculate the number of loosening degrees. For an M8 adjusting screw with a thread pitch of 1.25mm, if the required clearance is 0.20mm, the degrees of loosening will have to be:

(360° x 0.20)/1.25 = 58°, approximately. As each face on a hexagonal nut is 60°, you can use it as a point of reference.

Finishing the job

Once all the valve clearances have been adjusted:

◆ Put the rocker cover back on with a new gasket.
◆ Re-tighten the fastening bolts without over-tightening them to avoid deforming the rocker cover and its seat.

4 After having positioned the engine in the correct phase (see positioning table on page 126), slide the feeler gauge in between the valve stem and the rocker end. If the blade feels loose you will have to reduce the clearance. If the blade won't go in, you will have to increase it.

5 To modify the clearance, just loosen the lock nut on the adjusting screw with a ring spanner and adjust the screw with a screwdriver. The feeler gauge blade must slide without being forced and without any play. Re-tighten the lock nut while holding the adjusting screw with the screwdriver to keep it from moving out of adjustment. Once the lock nut is tightened, check the clearance adjustment.

6 Position the engine for a new phase and adjust the other valves in the same manner.

7 When the clearance on all the valves is set, thoroughly clean the gasket seat.

8 Inspect the gasket. If it is an O-ring and shows the least sign of cracking or deformation, change it. If the gasket is paper, replace it. Stick it on the rocker cover with grease or gasket goo. Place the cover back on the cylinder head, centring it carefully.

9 Tighten the fastening bolts, taking care not to squash the cover. Re-connect the breather.

VALVE CLEARANCE ADJUSTMENT

1-cylinder engine
Rocking method

Rock the valves on cylinder number:	Turn the engine 360°, ie one turn; adjust the valves on cylinder number:
1	1

2-cylinder engine
Rocking method

Rock the valves on cylinder number:	Turn the engine 360°, ie one turn; adjust the valves on cylinder number:
1 2	1 2

IMPORTANT

You must stick to the manufacturer's specifications for the clearance and adjustment method. If the valve clearance is too small, the valve opens too early and closes too late. Conversely, if the clearance is too large, the engine becomes noisy and wear is accelerated.

4-cylinder engine
Order of injection 1 3 4 2
Rocking method

Rock the valves on cylinder number:	Adjust the valves on cylinder number:
4	1
2	3
1	4
3	2

4-cylinder engine
'Wide open' adjustment method

Completely open the valves	Adjust the valves
E1 and I2	E4 and I3
I1 and E3	I4 and E2
I3 and E4	I2 and E1
E2 and I4	E3 and I1

Between each phase, turn the engine 180°, in its direction of rotation. I = intake; E = exhaust.

VALVE CLEARANCE ADJUSTMENT

3-cylinder engine

Phase 1	Place cylinder **1** at TDC, compression stroke	Adjust valves **I1**, **E1** and **E2**
Phase 2	Turn the engine 240°	Adjust valves **I2**, **I3** and **E3**

Between each phase, turn the motor 240°
in its direction of rotation

6-cylinder in line engine

Order of injection 1 4 2 6 3 5

Rocking method

Rock the valves of cylinder number:	Adjust the valves of cylinder number:
6	1
3	4
5	2
1	6
4	3
2	5

Setting the valve clearances without a feeler gauge

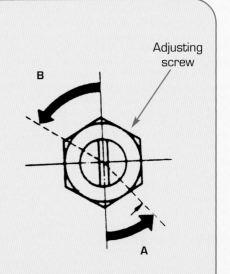

It is sometimes difficult, if not impossible to check the gap using a feeler gauge. If this is the case, calculate the number of degrees the screw should be loosened, taking into account the thread pitch of the adjusting screw.

Adjusting screw	**M8–1.25**mm
Degrees of loosening	**58°**

B = one side of the adjusting nut   **A**
360/6 = 60°

Set the clearance to O by tightening the adjusting screw without pushing the valve in. Then loosen it the number of degrees calculated.

Maintaining and adjusting a stuffing box

Simple

- 15 minutes
- Basic tools

Your vessel will be equipped with either a stern tube packing gland – the traditional stuffing box – or a rotary seal, or a lip seal. The installation and maintenance of these components require a lot of care and attention, particularly when launching or at each haul out, to ensure that this equipment is in perfect condition.

● When first launching (to be done on the water)

Loosen the stuffing box completely to fully flood the stern tube and wet the stuffing. Progressively and alternately re-tighten both clamping bolts (no more than one turn at a time); verify that both clamping plates are parallel.

The correct tightness is obtained when the shaft can still be turned freely by hand and the stuffing box lets in one drop of water every 10 to 30 seconds when the boat is motoring.

● During the season

Each time you go out, check the stuffing box drip rate and temperature, as well as the stuffing box cooling system if your boat has one.

During long inactive periods in the water, the stuffing box can be re-tightened, but when you start to use your boat again, you will have to go through the same procedure as when it was first launched.

For boats often out of the water (drying while at anchor or transported by road, for example) it would be wise to pay particular attention to the stuffing box and, if necessary, repeat the same procedure described for a first launch.

● Re-launching after winter storage

Change the stuffing before putting the boat back in the water. After launching, go through the same steps as for the first launch.

1 The stern drive unit with its traditional stuffing box. Although there are still a lot of boats where the stern tube seal on the propeller shaft is made by compressing three or four rings of stuffing around the shaft, this system is becoming rare. Today manufacturers usually install a rotary or lip seal.

2 When first launching, remember to loosen the stuffing box to flood it. If, even when loosened, no water seepage is apparent and the stuffing box heats up, unfasten the compression plate completely and slide it back along the shaft. Remove the stuffing rings one by one with the help of a hook until water seeps through. Then put the stuffing back in place.

3 You will notice that the bilge has been cleaned! The stuffing box bolts now have to be re-tightened simultaneously. You must be able to turn the propeller shaft by hand. It is normal to see water dripping. The correct tightness must allow a drop of water roughly every 30 seconds. If the stuffing box is almost at its maximum setting, and cannot be tightened any further, it will have to be re-packed.

Checking a rotary seal

Simple

- 5 minutes
- Basic tools

The rotary seal on the propeller shaft needs no maintenance but it should still be inspected periodically. No visible seepage is required, but it is essential to ensure that water reaches the seal to lubricate it.

◆ Periodic inspection and maintenance

The whole system will have to be inspected at least once a year, after long periods of disuse or when winterising.

To replace any of these parts, the boat will need to be hauled out.

At each launching or after each haul out, carefully draw the sealing boot back slightly. Ensure that water penetrates well inside it to lubricate the surfaces of the parts in contact.

◆ During the season

Each time you use the boat, inspect the rotary seal (clamp, boot), to make sure that all is well.

1 This generation of stuffing box, which no-longer uses stuffing, tends to achieve a complete seal. The seal is provided by two rings. One is fixed to the propeller shaft and the other is on a rubber boot. The elasticity of the boot maintains the two rings in contact under slight pressure. Adjusting the pressure of the propeller shaft ring on the boot ring provides the seal.

2 At each re-launching or after each haul out, remember to draw back the boot gently to expel any air trapped in the upper part of the stern tube.

Changing the stuffing box packing

Technical

- 1 hour
- Basic tools

Repacking the stuffing box will have to take place on the hard, every two or three seasons (or when the stuffing box plate is almost at its maximum setting).

Removing the stuffing

- Unfasten the stuffing box compression plate; slide it back.
- Empty the stuffing box with the help of a hook or a small screwdriver, without scratching the shaft or the box.
- Count the number of packing rings used.
- Degrease the stuffing box assembly with a brush and petrol.
- Lightly sand the shaft after having drawn back the compression plate.

> If the stuffing box is fitted with a greaser, empty, clean and re-pack it with new grease. Check that the grease gets through, then wipe off surplus grease with a clean rag.

Re-packing the stuffing box

- Re-pack the stuffing box by winding the new packing around the shaft to form separate rings of exactly the right circumference.

- Chamfer the ends of the packing in the tightening direction, so that they meet firmly together.
- Put on the new packing rings one after the other, taking care to offset the position of each joint to avoid any direct passage of seawater.
- Put the stuffing box plate back in place.
- Tighten the clamping bolts moderately (the compression sleeve must go in a distance equivalent to the thickness of one packing ring).

Adjustment

The tightness is correct when the disconnected shaft can still be turned by hand. If the stuffing sleeve goes completely into the stuffing box, add an extra packing ring. Conversely, if the sleeve doesn't go in enough, remove one ring.

Progressively perfect the seal adjustment once the boat is on the water and after 30 minutes under power. The setting is correct when the stuffing box lets in one drop of water every 10 to 30 seconds. Also check the stuffing box temperature by touching with your hand. It should be warm, never burning hot.

> The new packing materials, especially those made of Teflon or graphite, have good thermal conductivity which considerably reduces general wear and leaks.

1 Unfasten the compression plate. Note the quantity and colour of water in the bilge. Look for water and oil leaks. A good cleaning and a careful inspection are needed.

2 Slide the compression sleeve on to the shaft, then empty the stuffing box.

3 Remove the old packing with the help of a corkscrew or a small hook.

4 After thoroughly cleaning the bilge, clean and check the propeller shaft, the connecting hose and its clamps. Note the corrosion marks at the level of the packing rings. The shaft is etched.

5 Regarding the packing, two choices are possible: packing sold by the metre and packing ready-cut to the desired shaft diameter. The latter has the advantage of being ready to use.

6 Cut the packing to the shaft diameter and bevel it in the tightening direction with the help of a Stanley knife.

7 Pack the stuffing rings into the stuffing box. Be sure to offset the joints to prevent any direct passage of seawater.

8 Push the compression sleeve in place and re-tighten the bolts whilst turning the shaft by hand. The compression sleeve should go in a distance equivalent to the thickness of one stuffing ring. Now tighten it very lightly.

The adjustment is then made on the water in several stages. First, let it leak with the propeller shaft turning to expel the air trapped in the upper section of the stern tube. Then tighten the compression plate until there is one drop of water every 30 seconds. Verify by hand that the stuffing box is not heating up. It should be warm, never burning hot.

Replacing the cutless bearing

Technical

- 1 to 2 hours if removing the propeller
- Basic tools

The cutless bearing, mounted on the strut or the stern tube outlet, completes the drive shaft assembly. This type of bearing needs no maintenance. Seawater provides the lubrication. On the other hand, the bearing should be replaced as soon as the play on the shaft is more than 1.5mm.

How to replace the bearing

Two choices:

1 – Propeller shaft removed

- Remove the bearing's locking screws on the strut.
- Push out the bearing with the help of a tube or rod with a diameter slightly smaller than the bearing.
- Clean the bearing housing.
- Slide in the new bearing.
- Re-tighten the locking screws.

2 – Propeller shaft in

- Remove the propeller.
- Remove the bearing's locking screws.
- Push out the bearing with the help of a screwdriver or a section of tubing.
- Slide in the new bearing – soapy water will ease the way.
- Re-tighten the locking screws.
- Put the propeller back on.

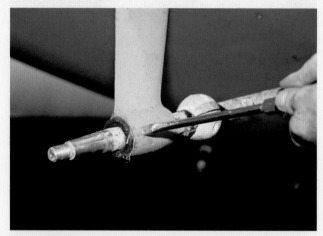

2 To change the bearing, the propeller needs to come off. Then unscrew the bearing's locking screws.

3 The most delicate step: remove the bearing. A section of tubing with an inside diameter equal to the shaft makes it easy to push the bearing out. A screwdriver may do the job as well.

1 Evaluate the play of the shaft in the bearing by hand. Replace the bearing when the play is more than 1.5mm.

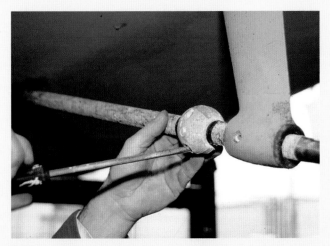

4 Remove the anode then tap out the bearing.

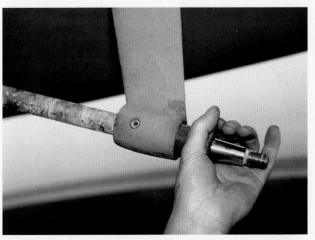

5 The worst is over; the bearing has been freed. You can remove it by hand.

6 Before putting in the new bearing, it is important to clean the shaft lightly with sandpaper.

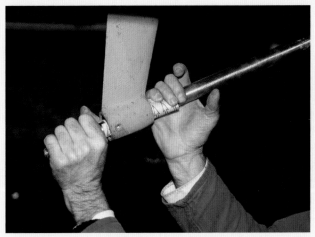

7 Also sandpaper the part of the shaft that fits in the bearing. No deposit must remain.

8 Put the new bearing back in with soapy water, making sure that the positioning holes will be aligned with the locking screws.

9 Screw the locking screws back in. Replace the anode and put the propeller back on.

Servicing the propeller shaft lip seal

Simple

- 5 minutes
- Basic tools

At each launch or after grounding

Remove the air from the joint by squeezing its sleeve and pushing it back on the propeller-end of the shaft. Squeezing the seal makes it open away from the shaft. All of the air has been expelled when water comes out of the gap created by pressing on the sleeve.

During the season

Follow the same procedure as for the rotary seal (page 63).

Periodic inspection and maintenance

Grease the seal with the manufacturer's recommended waterproof grease every 200 hours of operation, at every launch or once a year.

 IMPORTANT
Since it is the seal's internal lips that make it watertight, using any tool or foreign object other than those recommended by the manufacturer is risky.

Applying waterproof grease to the seal

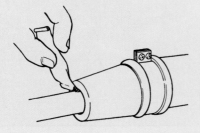

1 Because of its minimal maintenance, a lip seal is now often used. It is simple and quick to install and requires very little maintenance. It should be greased about every 200 hours and replaced after every 500 hours of operation or every 5 years.

2 At every launch or after grounding, remember to purge and lubricate the lip seal. Squeeze the sleeve, pushing it back toward the propeller-end of the shaft to expel any air trapped in the stern tube.

Aligning the propeller shaft

Technical

- 2 hours
- Basic tools, feeler gauge

All diesel engines vibrate to a certain extent even when correctly tuned. Although rubber engine mounts reduce this vibration, a defect in the propeller shaft alignment will increase it considerably and may cause damage to the gearbox.

Visual inspection of the alignment

Start by checking for any sagging or shearing of the rubber engine mounts by pushing down on the engine with a heavy duty lever. If the misalignment exceeds 5 microns per centimetre of coupling, re-align the engine and gearbox with the adjusting bolts on the engine mounts.

If the propeller shaft shakes when turning and if the problem persists or worsens once the nuts on the coupling have been re-tightened, the height adjustment or the shaft alignment in relation to the engine is no longer accurate. It is time to re-align the shaft.

Engine mounts

Depending on the boat type and construction, we find two types of mounts.

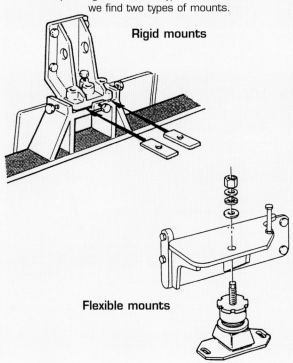

Rigid mounts

Flexible mounts

1 Unscrew and remove the nuts on the coupling.

2 Push the propeller shaft out and check that it is centred in the stern tube. Put it back in and engage the shaft coupling with the gearbox coupling, leaving a gap just small enough to slide a feeler gauge blade in.

3 At point A, insert the feeler gauge blade that will serve as reference. Example: 0.20mm feeler gauge blade.

Checking the alignment

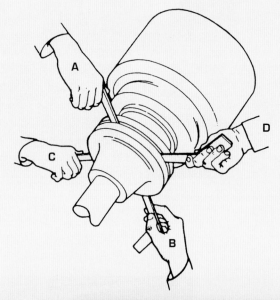

Height alignment
Adjust the brackets' height (see diagram). Screw or unscrew them a bit at a time and check the alignment of the couplings at each height variation.

Lateral alignment
Move the engine very slightly towards one side or the other. Once it is aligned, tighten the nuts and lock nuts on the brackets. If alignment is impossible, there are three probable causes: warped couplings, the propeller shaft is bent, the engine brackets are crooked.

4 Then, compare the dimensions at B, C and D. If the mis-alignment exceeds 5 microns per centimetre of coupling diameter, the engine and gearbox must be re-aligned with the propeller shaft.

Remember that the engine is aligned laterally in relation to the propeller shaft by unfastening the nuts on the engine mount lug bolts and the height is aligned by modifying the height of the fastening brackets in relation to the engine mounts.

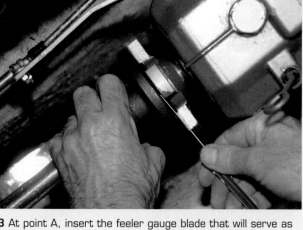

5 Loosen the nuts and lock nuts on the engine brackets.

6 After tightening the brackets and putting the bolts back on the coupling, check the alignment one last time.

Effect of engine bracket height on coupling adjustment

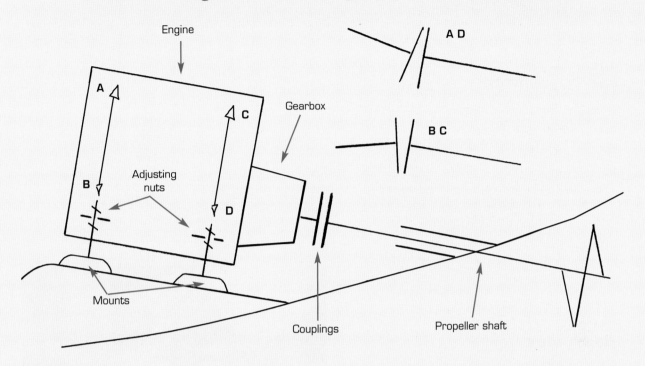

Effect of engine position on propeller shaft lateral alignment

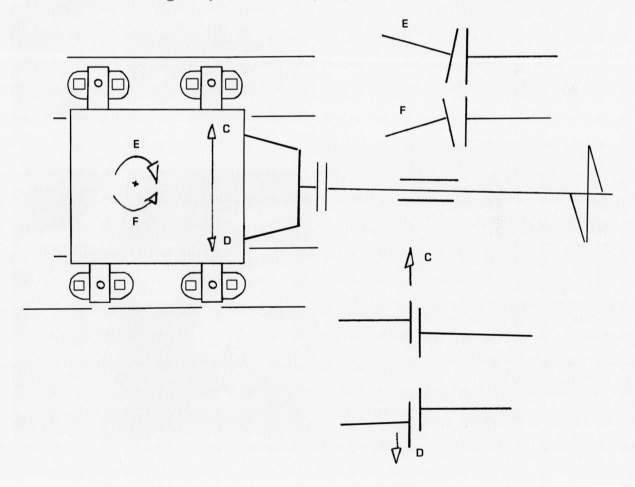

Removing the propeller

Technical

- 30 minutes
- Basic tools

A propeller needs no maintenance other than the periodic cleaning of the blades but if – when sailing – the propeller strikes a rock or some other obstruction, one or more of the blades could be seriously damaged and cause an imbalance and severe vibration in the drive shaft.

An incident of this kind requires a haul out to completely remove the propeller shaft, as well as a thorough inspection of the drive unit components such as the propeller, engine brackets and shaft.

Removing the propeller

- Unfasten the screw on the shaft end anode.
- Remove the anode.
- Bend back the tabs on the propeller nut lock washer.
- Unscrew and remove the nut (hold the propeller with the help of a wood block).
- Remove the propeller by hitting the propeller hub with a heavy hammer while pushing back on the opposite side (never hit the blades).
- Disengage the propeller.

 If resistance is excessive, use the appropriate extractor.

- Dislodge the woodruff (prop) key by tapping on its end with a cold chisel and a hammer. Inspect the key and the keyway. They must not show any sign of shearing or deformity.

Reassembling the propeller

Before putting the propeller back on, it has to be cleaned, along with the cone. To do this, use a scraper or a wire brush. Finish by polishing with an orbital sander.

! PRECAUTION
If the propeller is covered with antifouling, remember to protect yourself. The dust can be toxic and inhaling it can be very harmful.

Put the propeller back on its cone. When reassembling the propeller, make sure after tightening it to fold the lock washer tab over one face of the nut.

Remember to fit a new shaft end anode.

1 First, unscrew the anode with an Allen key.

2 A little tapping with a mallet is usually enough to loosen the anode.

3 Fold back the tab on the propeller nut lock washer with a hammer and cold chisel.

4 Jam the propeller with a large wooden block against the hull, then loosen the nut.

5 Freeing the propeller is an awkward operation. You will have to get someone to help. You will need a good hammer and a very heavy, large wooden block to push against. Now, hit the hub carefully but smartly.

6 Preferably use an extractor. Tension it and hit the end of the extractor hard.

7 Another hammer-and-chisel operation will be needed to free the key (with great care).

8 Clean the propeller. It's easy and quick if you use a rotary sander.

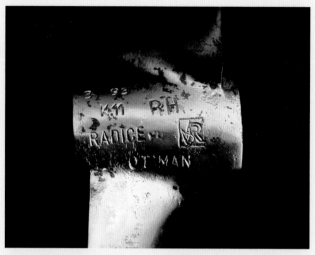

9–10 After a good clean, the propeller's manufacturer, rotation direction, diameter and pitch will be visible – all the information needed if you change the propeller.

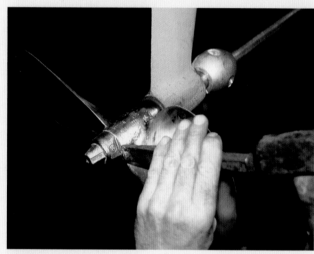

11 Replace the key after you have made sure it is in good condition. It must not be deformed or marked. Put the propeller back on. Screw the nut back on with a new lock washer, then tighten the propeller, 'blocking' it as you did during removal.

12 Fold over the lock washer tab.

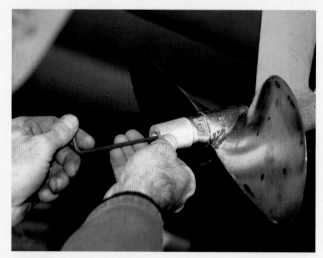

13 Screw a new anode on.

14 Job done.

Torque table for standard bolts

Use this table if you don't have the repair manual for your engine and don't know the manufacturer's recommended tightening torque.

● How to use it

Find out:

1 The quality of the bolt (R) in N/mm² (Newtons per square millimetre), which is a measure of a bolt's tensile strength.
2 Nominal diameter of the bolt to be tightened. In the table below, find the maximum recommended tightening torque.

Description	8.8		10.9		12.9	
Thread diameter (mm)	R ≥ 800 N/mm²		R ≥ 1000 N/mm²		R ≥ 1200 N/mm²	
	Nm	Kgm	Nm	Kgm	Nm	Kgm
4 x 0.70	3.6	0.37	5.1	0.52	6	0.62
5 x 0.80	7	0.72	9.9	1.01	11.9	1.22
6 x 1.00	12	1.23	17	1.73	20.4	2.08
7 x 1.00	19.8	2.02	27.8	2.84	33	3.40
8 x 1.25	29.6	3.02	41.6	4.25	50	5.10
9 x 1.25	38	3.88	53.4	5.45	64.2	6.55
10 x 1.50	52.5	5.36	73.8	7.54	88.7	9.05
12 x 1.75	89	9.09	125	12.8	150	15.30
14 x 2.00	135	13.80	190	19.40	228	23.3
16 x 2.00	205	21.00	289	29.50	347	35.40
18 x 2.50	257	26.30	362	37.00	435	44.40
20 x 2.50	358	36.60	504	51.50	605	61.80
22 x 2.50	435	44.40	611	62.40	734	74.90
24 x 3.00	557	56.90	784	80.00	940	96.00

REPAIRS

EVEN THOUGH SOME MAINTENANCE JOBS are fairly easy, reassembling an engine and making it work requires a level of technical ability that not everyone has, and so the success (or otherwise) of a repair may depend to a large extent upon how well the trouble shooting procedures and manufacturer's recommendations have been followed. This chapter is not intended to be a substitute for your engine repair manual. Rather, it is a summary of the various necessary inspections and precautions you need to take to successfully dismantle and reassemble your engine.

Removing the cylinder head

Complex

- 2 hours
- Basic tools

A leaking head gasket is a relatively common problem with older engines. The following are usually signs of the problem:

- Loss of power.
- Indirect cooling systems: water consumption and bubbles coming to the water surface in the exchanger.
- Emulsified water and oil in the sump. The oil has the consistency of mayonnaise.
- The engine overheats.

Even though the engine will continue to run the problem will undoubtedly become worse so it is time to remove the head and change the gasket.

Depending on the engine design: rocker arm (Yanmar, GM or Volvo 2000) or overhead cam (Perkins, Prima, Perama, and Volvo 22) – the task will be slightly different.

To remove the head on overhead cam engines, you will first need to remove the timing belt that drives the camshaft.

Removing the head

This is not particularly difficult. But note: the head should only be removed when the engine is cold to avoid deforming it.

How to do it

- Close the seacock.
- Isolate the power.
- Drain the salt water circuit.
- Drain the engine coolant from its lowest point.
- Remove or disconnect the exhaust pipe.
- Disassemble the water cooling system lines.
- Dismantle the high pressure injection lines.
- Dismantle the injector return line.

Overhead cam engines

- Remove the timing belt.
- Disconnect the water temperature sensor.
- On engines with indirect injection, disconnect the glow plugs.
- Remove the cylinder head cover.

Rocker arm engines

- Remove the rocker arm assembly and if needed, the oil lines.
- Remove the push rods.
- Progressively loosen the engine head bolts or nuts a bit at a time.

1 First remove the peripherals: exhaust manifold and inlet manifold.

2 To remove the threaded rod, tighten two nuts, one against the other, then unscrew the first one.

Generally, the head bolts or nuts are loosened in inverse order of the manufacturer's instructions for tightening them. In other words, first loosen the ones farthest away from the centre of the head. Continue loosening them in a criss-cross fashion to progressively release the tension.

Remove the head. Avoid resting it on its face.

> ### ! IMPORTANT
> **Never stick a screwdriver into the joint between the head and the block to pry the head off the block. Instead, use the bosses that are on the sides of the block and head.**
>
> **If, despite everything, you still can't free the head, crank the starter motor and the head should come loose.**
>
> **On some engines the head is bolted on with lug studs. If you need more clearance to get the head off, you will need to remove the lug studs as well.**
>
> **Before removing the head, prepare a clean place to set it down.**

● Checking and inspecting the head

Once the head is removed carefully check the head and engine block faces. Also check the old gasket. You can often find the spot with the leak by carefully looking at the gasket crimping. Dark marks are the sign of a leak. Clean the head.

Inspection
Inspection will tell you about the following:

◆ Warping of the faces (head, engine block)
◆ Combustion chamber condition (there may be cracks)
◆ Deposits in the water passages.
◆ If the head is warped it must be taken to be resurfaced at a specialised workshop. Each manufacturer indicates the maximum allowable warping.
◆ Check the condition of the valves. If you have any doubts about their seals, they must be dismantled. (See pages 148–55.)

3 Remove the rocker cover. If it is stuck, use a wooden mallet.

4 To get at the head bolts, you must remove the rocker arm assembly. Note how the rocker arms are designed. They can be in one single assembly, as shown, which makes dismantling easy, or they can be bolted individually onto bosses on the head.

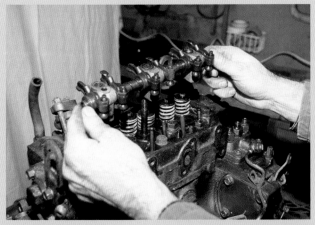

5 When all the bolts or nuts holding the rocker arm assembly have been taken out, remove it. Carefully note the location of each bolt as it is removed so it can be returned to its original place.

6 On Yanmar engines, remember to save the valve stem caps.

7 Remove the push rods one at a time.

8 Loosen the head bolts, starting with the ones furthest away from the centre. These bolts or nuts are very tight. Use a socket with a square fitting. If necessary, use a pipe for an extension.

9 Remove the head. If it is stuck fast, never pry it off with a screwdriver in the joint surface. It is best to use leverage on the bosses.

10 Look for signs of leaks on the gasket. A black mark on the crimping that shows residues of combustion is the sign of a bad seal.

Replacing the head gasket

Complex

- 2 hours
- Basic tools

Ensure that everything is clean. Make sure that you have a new gasket or gasket kit; the head has been cleaned, air blown, inspected and checked. You have the tools, the torque wrench and the torque specifications. You are ready to refit the head.

● To refit the head

Put the new gasket in place. Check that it is positioned properly. Most head gaskets have an inscription or mark that says 'top', which indicates that this side should face up, ie towards the head. Also check that the openings in the cylinder block for water and oil match.

Some manufacturers sell head gasket kits. These kits have all the seals needed for the parts on the head, eg the exhaust manifold, heat exchanger, water pump seal, inlet manifold, cylinder head cover, etc.

Position and tighten the head.

> **! IMPORTANT**
> **With an overhead cam engine (camshaft on the head), before remounting the head, turn the crankshaft a few degrees to position the pistons at mid-stroke. This is to prevent the valves and pistons from colliding.**

● Precautions

Follow the manufacturer's instructions on the tightening sequence and the amount of torque for the head. You *must* use a torque wrench to do this.

In general, you tighten from the centre out to the ends. If the engine has several heads, make sure that the exhaust manifold is aligned with them to avoid warping and deformations.

An overhead cam engine
A timing adjustment must be made. Refer to the worksheet 'Changing the timing belt', page 176.

A rocker arm engine
The valve clearances must be adjusted. Refer to the worksheet 'Adjusting the valve clearances', page 122.

Refit the head using new seals and reconnect the parts to the head: cylinder head cover, exhaust pipe, water temperature sensor, high pressure lines, injector overflow line, etc.

Refill the cooling system with new 4-season coolant.

> **! IMPORTANT**
> **If coolant has contaminated the sump, drain the oil after refitting the head. Flush out and refill with the correct grade of oil.**

If the engine doesn't start right away, the high pressure lines have to be bled by unscrewing the lines at the injector one quarter of a turn. See the worksheet 'Bleeding the fuel system', page 99.

● Take the following precautions:

Remember that you have to re-tighten the head once the engine has run for about 10 hours. This must be done when the engine is cold (ie stopped for several hours). Once you have done this, you should also check and, if need be, adjust the valves.

1 Position the new gasket. Make sure it is oriented correctly. The mark 'top' should be facing up (toward the head). Check that the pattern corresponds exactly to the holes, and water and oil openings.

2 Before fitting the new gasket you must clean the block and head faces scrupulously. Every surface on the block and head has to be dust and grease free. Refit the head.

3 Oil the head bolts or nuts, then tighten them one-third of the specified torque in the order indicated by the manufacturer. Next, tighten them two-thirds of the specified torque and finally to the manufacturer's recommended torque. Check with the workshop manual to see whether or not the head bolts then need to be angle-tightened beyond the basic recommended torque.

4 Re-insert the push rods in their places. Put the caps on the valve stems (Yanmar). Remount the rocker arm assembly. Tighten its bolts progressively. Here too, use the correct torque.

5 On rocker arm engines, remember to readjust the valves after you tighten the head and rocker arm assembly.

6 Reassemble everything in reverse order to disassembly. Grease the injectors with graphite grease before putting them back in the head.

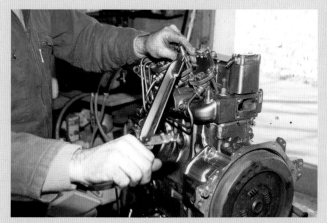

7 Tighten the injector hold-down plate to the manufacturer's recommended torque.

8 Refit the rocker cover using a new gasket. You may want to apply some gasket compound or grease to the gasket to hold it in place.

9 Refit the intake and exhaust manifolds using new gaskets.

Angle of tightening

Tightening with a torque wrench followed by tightening to an angle

Some manufacturers recommend a two step tightening method:

◆ Step 1: The assembly is tightened to the manufacturer's specified torque.
◆ Step 2: Tighten again to a specific angle.

Example

For a tightening angle of 90°, the assembly will be tightened an extra quarter of a turn after tightening to the prescribed torque.

For assemblies where no tightening torque is given, it can be done by following the table below. The values shown are approximate.

Tightening torque

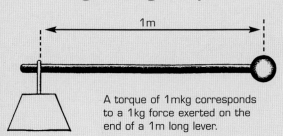

A torque of 1mkg corresponds to a 1kg force exerted on the end of a 1m long lever.

The dimensions of the flat or ring spanners are designed so that normal human strength can correctly tighten a nut or bolt. It is a question of leverage. You have probably noticed that an 8mm spanner is much smaller than a 13mm.

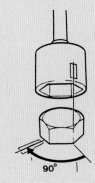

Designation	Torque: Nm
M5	5
M6	10
M8	20
M10	40
M12	70
M14	115

The nuts and bolts are divided into different classes of resistance, indicated by a mark on the head of the bolt. The higher the number, the greater the resistance.

Example

A 10-9 bolt is stronger than an 8-8 bolt. It is easy to see why it is important to put the bolts back in their original place when remounting an assembly.

Reconditioning the cylinder head

Complex

- 4 hours
- Valve-spring compressor, calipers, straight edge, feeler gauges

If you notice that your engine is losing power or that it is hard to start, compression loss is probably the cause. However, the signs can sometimes be even more alarming. A complete loss of compression in one cylinder may indicate a burnt valve. In any case, it is important to remove the head to check the head and block faces for warping; the valves, valve seats and valve faces.

If you haven't removed the head yet and you are unsure whether or not you need to do it, checking the engine compression will help you decide (refer to the worksheet 'Testing the compression', page 206).

● Dismantling the head

To check and recondition the head, it must be completely dismantled and cleaned.

- Remove the inlet manifold, the exhaust, and in the case of indirect cooling, the heat exchanger.
- Remove the injectors.
- Depending on the engine type, remove the cylinder head cover.
- Remove the valves.
- This job requires the use of a valve-spring compressor.
- Begin by tapping the edges of the valve spring retainers with a mallet to free the valve keepers.
- Compress the springs with the valve-spring compressor and remove the valve keepers with needle-nose pliers.
- Release the springs.
- Remove in order:

  1 the valve spring retainer
  2 the spring(s)
  3 the lower cup
  4 the base washer (found only on alloy heads)
  5 the valve

- Repeat for each valve. Note the position of each part.
- Remove the valve stem oil seals from the valve stems.

> **! IMPORTANT**
> The valves are not interchangeable. Each valve has its own spring, spring retainer and keeper.

● Check and inspect

Check the **valve stem play** in the valve guide (a few hundredths of a mm, maximum). If there is too much play, the valve guides must be changed. This must be done before grinding or replacing the valve seats.

Then check the **valve margin**. Too small a margin can cause difficult starting and exhaust smoke. Too large a margin allows the valve to hit the piston.

Check the valve faces, the seats and the valves themselves:

- Lightly pitted: simple grinding-in
- Lightly hollowed: resurface
- Hollowed head: replace

Check the **width of the valve faces**. It should not exceed 1 to 1.2mm. If it is larger, the valve must be resurfaced.

Check the **springs' condition and length**. Be careful: the intake and exhaust springs are different. Don't mix them up when removing them.

Check the **head and block faces** for warping with a straight edge. Measure the amount of warping by sliding a feeler gauge between the sealing surface and the straight edge at every point, as indicated in the drawing.

If the warping exceeds the maximum allowed by the manufacturer, the head has to be resurfaced. Generally, no more than 0.05mm of warping is allowed for a head 500mm long.

Measure the **thickness of the head**. If the thickness of the head after resurfacing is less than the manufacturer's specification, it will have to be replaced.

● Other points to check

A diesel engine cylinder head has injector housings. Pay particular attention to these. Check the rings and systematically change the copper seals.

Also, any serious head overhaul has to be tested before reassembling the valves. The purpose of the test is to check tightness of the head and is done by a resurfacing workshop.

● Grinding-in the valves

This consists of rotating the valves in their seats using a valve grinding tool. First, de-scale the valves and their faces with a scraper and wire brush.

Closely examine the valves during cleaning. Reject any that are cracked, pitted, bent or too worn at the face or the stem. Small pits or burns on the faces will disappear when the valves are ground in.

Make sure you don't mix up the valves during cleaning. Each one must be put back in its original place. Coat the valve face with abrasive paste, making sure you do not get it on the stem. Place the valve in its seat.

Stick the grinding tool's suction cup to the valve head. Spin the grinding tool back and forth while pressing down heavily on the valve.

After doing this about ten times, lift the valve and turn it $1/4$ of a turn. When you have done this several times, clean the valve seat and coat it with new abrasive paste. Repeat the operation until all traces of burns or pitting have disappeared. After grinding in and cleaning, the face should be a uniform dull grey and should not be wider than 1.2mm.

Clean the valves and seats with petrol and a brush. Blow dry with compressed air.

 Unless you have new valves and seats, always begin grinding with a coarse abrasive paste and finish with a fine one.

● Refitting the valves

◆ Oil the stems and seats with fresh engine oil.
◆ Refit the valves making sure that the ones you have ground are put back in the same seat.
◆ Refit each spring and retainer in the reverse order to the order you removed them.
◆ Compress the springs with a valve-spring compressor.
◆ Replace the keepers in their grooves.
◆ Release the spring.
◆ Tap the valve stems with a mallet to make sure that the keepers are in place.

 IMPORTANT
Before reassembling the valves, remember to change the valve stem seals on the valve guides.

● Reassembling the head

◆ Refit the exhaust and intake manifolds, and in the case of indirect cooling systems, the heat exchanger.
◆ Replace the injectors, remembering the copper seals.
◆ Refit the rocker arm assembly, or the push rods in the case of an overhead cam engine. In this case, you can then adjust the valve clearances, before remounting the head on the engine.

● Refitting the head

Refit the head using a new gasket, following the manufacturer's recommended bolt tightening sequence. See the worksheet 'Replacing the head gasket', page 145.

Rocker arm cylinder head

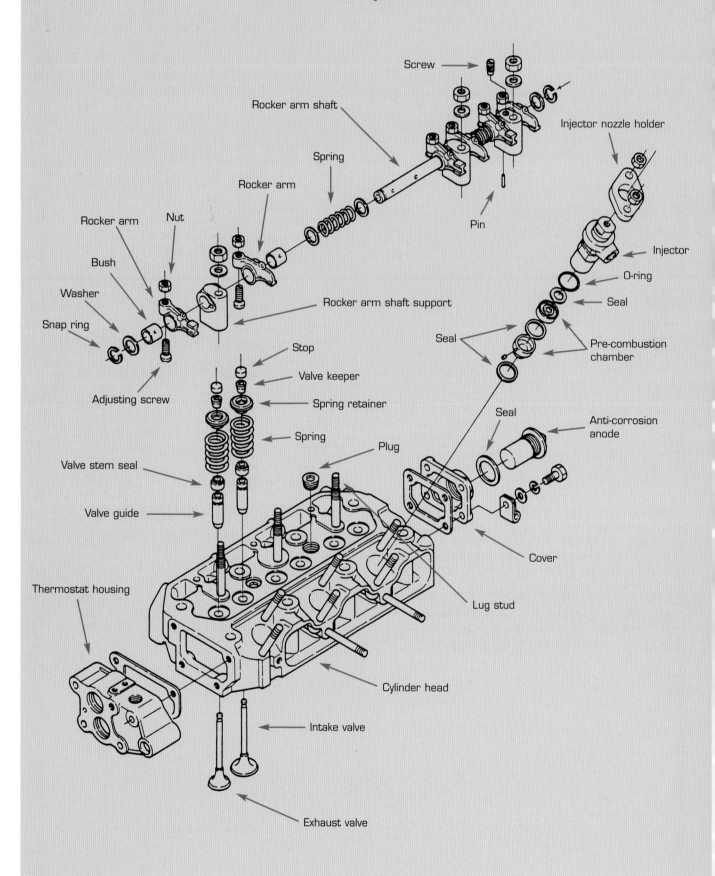

Screw

Rocker arm shaft

Spring

Rocker arm

Rocker arm

Nut

Bush

Washer

Snap ring

Adjusting screw

Rocker arm shaft support

Pin

Injector nozzle holder

Injector

O-ring

Seal

Pre-combustion chamber

Seal

Stop

Valve keeper

Spring retainer

Spring

Plug

Valve stem seal

Valve guide

Seal

Anti-corrosion anode

Lug stud

Cover

Thermostat housing

Cylinder head

Intake valve

Exhaust valve

1 Use a valve-spring compressor to remove the valves. First tap the edges of the spring retainers lightly with a mallet to release the valve keepers.

2 Compress the spring, then remove the keepers, making sure that the spring doesn't release. Be careful not to lose the keepers.

3 After releasing the spring, remove the spring retainer, spring, and valve spring washer.

4 Set aside the valve retainer, spring, and washer assembly in a way that you can re-install them in the same order. Here, the valves are set on a piece of wood with 8 holes, with 'cylinder 1' (next to the flywheel) marked with an arrow.

5 Even if you don't have a valve-spring compressor, it is still possible to do the job by putting a pipe on the spring retainer and hitting it with a large hammer.

6 This spring, suddenly compressed, will free the valve keepers. To avoid damaging the head gasket surface, place the head on a workbench or a wooden stand.

7 All of the valves have been removed. Now they have to be carefully cleaned before they can be examined.

8 Remove the valve stem seal by levering it with a screwdriver.

9 This seal, which is mounted on the upper part of the valve guide, prevents oil being sucked in around the guide and valve stem during the intake cycle. It reduces oil consumption and for this reason, it should be replaced each time it is removed.

10 The valve stem and guide are both subject to wear. To check the degree of wear, put the valve in its guide, allowing it to stay open and protrude about 1cm. If you can move the valve from side to side or feel any play, the valve stem or guide are worn. Repeat this with a new valve. If the play has disappeared, the valve stem is worn. If it hasn't, the valve guide is worn.

Checking for valve-guide wear

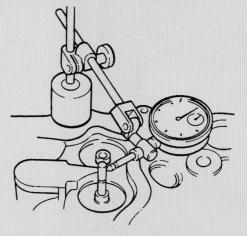

Check valve-guide wear with a dial indicator.
Maximum play = 0.15mm.

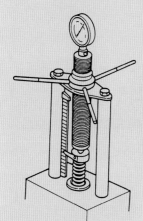

11 It is common for springs to sag. Measure the spring length and compare it to the manufacturer's specification.

Check the valve springs

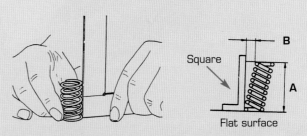

Square

Flat surface

Also check the spring for squareness and how it seats. When it passes the wear limit **B**, it causes the valve guide to become oval in shape. Check the length A.

Check the spring tension

The spring tension can be checked with a special tool. If it is less than the specification, it has to be changed.

12 Check the head for warping. The face has to be perfectly clean. Head warping damages the head gasket and causes compression leaks. Try to slide a feeler gauge between the gasket sealing face and the straight edge.

Check the cylinder head and engine block faces for warping

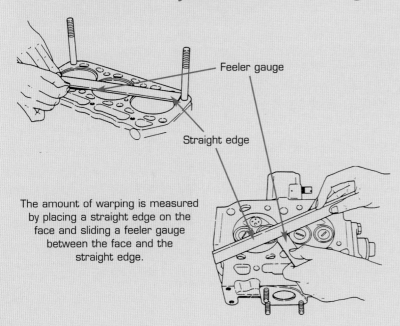

Feeler gauge

Straight edge

The amount of warping is measured by placing a straight edge on the face and sliding a feeler gauge between the face and the straight edge.

The measurement should be taken at the four corners and the two diagonals. Use the largest measurement.

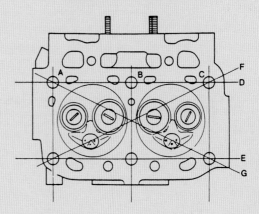

13 Also measure the head's thickness. Compare this measurement to the manufacturer's specification to find out if it has already been resurfaced.

14 To grind a valve, you need to coat its face with abrasive paste. You should start with a coarse grain before using a fine one.

15 After returning the valve to its place, stick the valve-grinding tool's suction cup on the valve and then rotate it back and forth with your hands. The grinding noise will be very harsh at first, but will become smoother after a few dozen back and forth rotations. Then raise the valve slightly, turn it $1/4$ of a turn and grind it again.

When you have done this several times, remove the valve and clean the face and seat. Repeat the whole operation several times until black areas and little burnt or corroded spots have disappeared.

16 After cleaning, the face and seat surfaces should be a uniform, dull grey. There should not be any traces of corrosion or pitting.

Valve seats

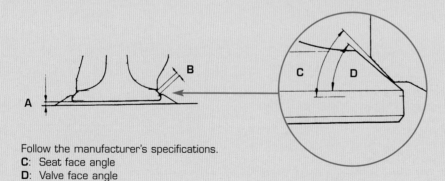

Follow the manufacturer's specifications.
C: Seat face angle
D: Valve face angle

If the valve seats are pitted or worn, they will have to be countersunk or re-bored. Check the face width. It should be less than 1mm. Check the valve margin and compare it to the manufacturer's specification.

17 At this stage of corrosion and wear, the valves should be changed and the seats on the head should be reground.

18 Remember to put on new valve stem seals.

19 Oil the valve stems with fresh engine oil. Carefully clean the valves before putting them back in their guides. There shouldn't be any traces of abrasive paste.

20 Replace the valves, making sure that each one matches the seat that has been ground. Replace the springs, retainers and keepers with the help of the valve-spring compressor.

21 Make sure that the keepers are properly positioned in their grooves. Then tap the valve stem end with a mallet to make sure the keepers are well set.

Dismantling the engine

Complex

● 3 hours

● Basic tool kit, piston ring expander

Many repairs require the engine to be almost completely dismantled. This is the case when rebuilding the connecting rod assembly or replacing the rings.

This worksheet will tell you how to do this, as well as the precautions to take if you are dismantling the connecting rod assembly.

● How to do it

◆ First, drain the engine oil.
◆ If you don't have to dismantle the crankshaft, you won't need to disconnect the gearbox from the engine.
◆ Remove the cylinder head.
◆ If it is an overhead cam engine driven by a timing belt, remove the timing belt's cover, loosen the cam belt tensioner, then remove the belt.
◆ Remove the lower plate or sump, giving access to the con-rod 'big end' bolts.
◆ Lay the engine on its side.

 If you pay close attention as you dismantle the engine, you may find answers to some of your trouble-shooting questions.

● Marking the components

Before removing or dismantling the piston/con-rod assembly, it is vital to mark all the parts to ensure that they are returned to their original positions. If you intend to re-use the pistons they must be returned to the same cylinder that they came from. The big end bearing caps must also be reunited with the original con-rod.

In most cases, the manufacturer has marked the parts or numbered them so you can match them correctly.

If your engine doesn't have any markings, it is very important to mark the parts before dismantling them. The simplest way is to stamp them, using a hammer and punch. For example, rod 1 is usually nearest to the flywheel. Working from the flywheel back, punch one dot on the con-rod cap and one dot on the rod itself. Always make the mark on the camshaft side so you will know which way to orient the parts when you replace them. Do the same for rod 2, and so on.

● Removing the con-rods and pistons

◆ Remove the nuts on the big end caps. This is best done with the respective cylinder at bottom dead centre (BDC).
◆ Remove the con-rod cap.
◆ Save the bearing inserts (shells).

> **! IMPORTANT**
> **Do not mix up the con-rod bearing shells. They each have to be refitted to their respective rods.**

◆ Before extracting the con-rod assembly, make sure you remove any carbon deposits from the upper part of the cylinder.
◆ Knock the con-rod assembly out with a bronze rod and mallet. On most engines, the con-rod assembly is removed through the top (towards the head). Be careful not to scratch the crankshaft journals with the connecting rod bolt threads.
◆ Turn the crankshaft to get to the next cylinder and bring its piston to BDC. Place the pistons on the bench in the same order they were removed from the engine.

● Removing the piston's con-rod

◆ Using a small screwdriver or needle nose pliers, remove the snap rings holding the piston gudgeon pins.

> **! IMPORTANT**
> **Never use a vice to hold a piston.**

◆ Separate the piston from the rod by removing the pin.
◆ For press-assembled pistons, use a special tool and a press.
◆ Open and remove the piston rings.

 A piston ring expander makes the task easier.

1 Once the sump is removed, you have easy access to the lower part of the engine.

2 Note the deposits in the bottom of the sump.

3 Loosen, then remove the big end cap nuts. To facilitate this, put the cylinder at bottom dead centre.

4 Detach and remove the con-rod caps. A light tap on either side with a hammer will loosen them.

5 Then knock the piston/con-rod assembly out through the top of the cylinder.

6 Note the order of assembly, tab against tab.

7 The con-rod caps and con-rods should be marked by the manufacturer. If not, mark them yourself.

8 All of the piston/con-rod assemblies are removed. Clean everything perfectly before checking, examining and measuring them.

Checking the piston/con-rod/cylinder assembly

Complex

⬡ 2 hours

⬡ Basic tool kit, ring-removing-and-replacing tool, feeler gauge, micrometer, calipers

After removing the piston/con-rod/cylinder assembly it can be disassembled, checked and rebuilt by an amateur provided he has some technical knowledge and a complete set of tools.

⬡ Cleaning the pistons

The pistons and piston ring grooves must be carefully cleaned. Do not use a wire brush or sharp scraper.

The deposits can be removed with a wire brush and petrol. For stubborn marks, use a Scotch Brite scouring pad. The piston ring grooves can be cleaned with a special tool. If you don't have one and if you're not reassembling the same rings, you can cut a piece of wood to fit the grooves and ream them out with that. Remember to clean the oil holes.

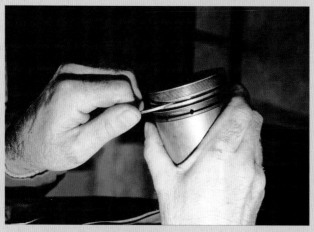

1 Clean the ring grooves.

2 Carefully clean the pistons. Check the play of the rings in their grooves with a feeler gauge.

Measure the wear in the piston ring groove

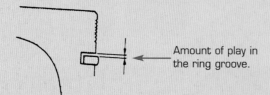

Amount of play in the ring groove.

3 Fit the ring in the cylinder below the worn area to check the end gap.

Examining the pistons

Carefully examine each piston. Reject any that have signs of wear or scratches at the head as well as at the skirt. If you intend to re-use them, check their diameter against the manufacturer's specified measurement. Also check the piston pin bearings. Change them if they are worn. The gudgeon pin must be replaced if it shows any signs of pitting or wear.

Checking the piston rings

Check the play of the rings in their grooves with a feeler gauge. The rings should be free in the grooves. If the gap measured is greater than that specified by the manufacturer (0.15mm on average), replace the rings or the piston/ring assembly.

Place the rings in the cylinder and check their end gap. In the absence of any instructions, an average value is acceptable:

0.30mm to 0.45mm end gap for up to 90mm bore
0.35mm to 0.55mm end gap for up to 100mm bore
0.40mm to 0.60mm end gap for up to 110mm bore
0.45mm to 0.65mm end gap for up to 140mm bore

4 Use a feeler gauge to measure the end gap.

Measuring the piston

0.5mm

1.0mm

0.26mm

60°

Details of the heat dissipation grooves

Groove measurement for the first compression ring.

Groove measurement for the second compression ring.

Groove measurement for the oil control ring.

Each measurement is important. If the wear limit is exceeded, the piston must be replaced.

Examining the piston rods

First make sure that the rod is not bent. Check or have a professional service engineer check the rod for squareness and torsion.

> ★ **An engine that is difficult to start or that has poor compression in one or several cylinders may have one, or several, bent rods. Generally, this problem, which is not uncommon, is due to water getting into the exhaust. (Water can get into the exhaust when motor sailing heeled over.)**

When changing a rod, make sure it meets the manufacturer's recommended weight tolerance.

When you remove the piston/rod assembly you do not always have to change the bearings. Before replacing the rod in the engine, carefully clean the journals as well as the oil passages.

5 The straightness of the rod can be quickly checked by laying it on a flat surface.

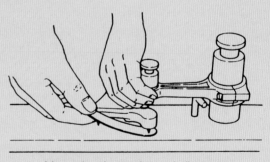

6 Here, no measuring tool is needed to prove that the rod is definitely bent.

Checking the con-rods

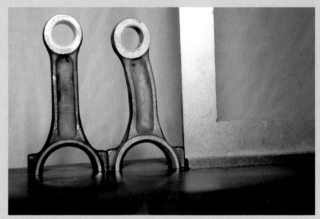

100mm

100mm

Alignment

Torsion

Poor alignment or torsion causes characteristic wear on the bearings and in the cylinders. When the measurements exceed the specified limits, the rod must be replaced.

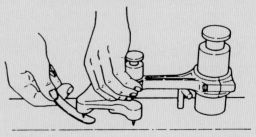

Measuring the torsion and alignment.

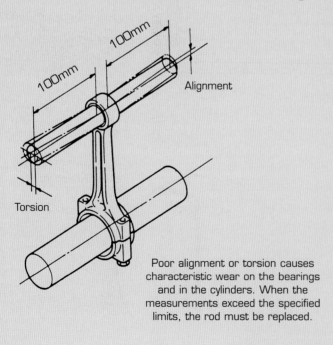

◆ Checking the crankshaft journals

Check the condition of the journal surface. No scratches are acceptable. With the aid of a micrometer, measure the journal's out-of-round wear and taper. Compare the measurements to the manufacturer's specifications.

If there are any deep scratches or deformities, the crankshaft will need to be completely removed so the journals can be resurfaced. When remounting the journals after resurfacing, be sure to use the correct sized bearing shells.

◆ Checking the bearings

If the bearings show traces of wear, scratches or incrustations, they must be replaced.

> **! IMPORTANT**
> **Before ordering new bearings, check their size. The crankshaft may already have been resurfaced. If so, the back of the bearing will be marked with the resurfacing measurement: standard bearing (std) and 0.10–0.20 when resurfaced.**

7 Taking measurements at diametrically opposed points allows you to verify if the journal is out of round.

8 Checking the bearings reveals a pink layer of metal, a sign of significant wear.

9 The bearing markings are on the outside of the shell. Std for standard size or 0.10–0.20 or the resurfaced size.

Diagram of hydrodynamic lubrication

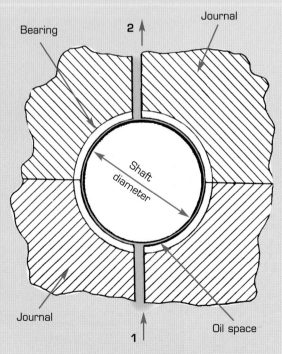

Bearing

2 ↑

Journal

Shaft diameter

Journal

1 ↑

Oil space

Each element to be greased has an inlet for pressurised oil **1** and an outlet leading to other elements on the circuit to be lubricated **2**.

Construction of a multiple-layer bearing

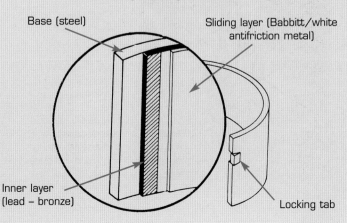

Base (steel)

Sliding layer (Babbitt/white antifriction metal)

Inner layer (lead – bronze)

Locking tab

● Checking the cylinder(s)

Clean the cylinder block completely before beginning any of these inspections and measurements. Make sure the cylinder interior is perfectly clean before inspecting the surface. The cylinder should be smooth and free from any scratches or dark stains.

When the engine runs, the cylinder walls are worn by the combination of piston movement, high temperature and pressure. The upper part of the cylinder is subjected to the greatest stress. It is in this area that the greatest wear is found. When the piston goes down, the pressure and temperature diminish and so does the wear.

The cylinder goes 'out of round' or ovalised as a result of the piston's lateral push on the rod.

If you don't have the correct tools, only a resurfacing workshop will be able to check if the cylinders are out of round or tapered. If you can borrow the tool, measure the sides of the bore along its two axes at three different levels. If the irregularities found are significant, ie larger than the manufacturer's specifications, the cylinder will have to be re-machined or re-sleeved. These two jobs must be done by a resurfacing workshop.

11 Honing removes small scratches and de-glazes the cylinders.

12 The cylinders should always be cleaned after resurfacing.

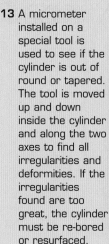

10 If there are scratches or dark stains on the cylinder(s), they will have to be resurfaced with a drill-mounted honing tool. If the wear or scratches are faint, honing is sufficient. Otherwise, the cylinders will have to be re-bored.

13 A micrometer installed on a special tool is used to see if the cylinder is out of round or tapered. The tool is moved up and down inside the cylinder and along the two axes to find all irregularities and deformities. If the irregularities found are too great, the cylinder must be re-bored or resurfaced.

Checking the cylinders

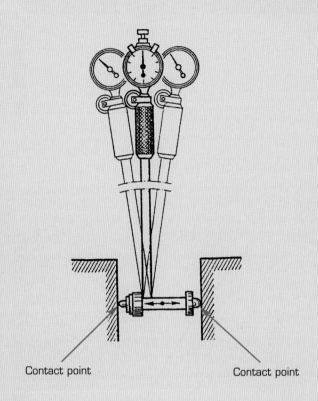

Contact point   Contact point

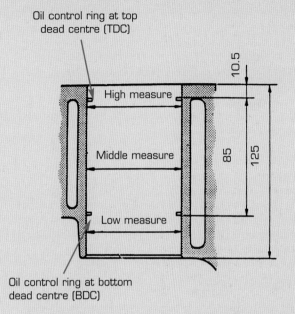

Oil control ring at top dead centre (TDC)

High measure

Middle measure

Low measure

Oil control ring at bottom dead centre (BDC)

The objective of the measurements you will take is to determine the amount of deformation or wear in the cylinder. They are made at three points in the cylinder: high, middle, low.

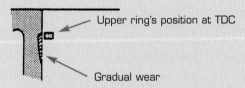

Upper ring's position at TDC

Gradual wear

When the piston is at TDC (at the beginning of the 3rd stroke Combustion/Expansion), the pressure in the cylinder is high. The rings push very hard on the upper part of the cylinder. It is also at this moment that the temperature is highest and the oil loses a some of its protective qualities. The result is characteristic wear on the upper part of the cylinder.

Measure the cylinder along two axes to see if it is out of round.

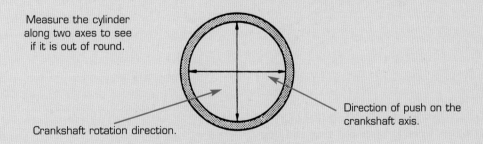

Direction of push on the crankshaft axis.

Crankshaft rotation direction.

The lateral push of the piston on the rod causes the cylinder to go gradually out of round.

Reassembling the engine

Complex

- 3–8 hours depending on the type of engine
- Basic tool kit, ring-removing-and-replacing tool, torque wrench

Having cleaned the parts and mating surfaces, checked each component and changed the ones that are worn, you are ready to reassemble the engine.

Reassembly is done in reverse order of disassembly. As you reassemble the components, make sure that all of them are perfectly clean and pay close attention to their exact position.

◆ Reassembling the piston rings

When reassembling the rings, pay particular attention to the direction in which they must be mounted. Always place the mark 'top' toward the top of the piston. They are put in place with a ring-removal-and-replacement tool. Start with the oil control ring and finish with the top compression ring.

◆ Bearing reassembly

Position the bearings and push on the con-rod heads and caps, making sure that nothing sticks out on either side. Caution: some caps have a lubrication hole. It is important to check that it coincides with the oil passage hole.

Oil the bearings with new engine oil.

◆ Positioning the rings and preparing them for reassembly

Oil the piston rings then position them so that the end gaps are staggered. This consists of positioning the ring end gaps along three different axes so they are never in line with one another.

For a piston with three rings, the end gaps should be set at 60° to one another. The oil control ring must be at 90° to the piston axis. Oil the interior of the piston ring clamp.

Tighten it to compress the rings.

> **! IMPORTANT**
> When you tighten the piston ring clamp, make sure you don't alter the ring end gap positions.

◆ Piston/rod reassembly

Caution: Each piston must be put back in the correct cylinder and in the right direction. Most pistons have either a notch or mark on the head, which must always face the front of the engine.

- Put the crankshaft at BDC for the cylinder in question.
- Oil the cylinder.
- Fit in the piston. Push it in by tapping the piston head with the handle of a hammer. Make sure the piston rod bearing faces the journal as you do this.
- Position the rod head on the journal.
- Put the corresponding piston rod cap on in the correct direction.
- Tighten the piston head nuts with a torque wrench to the manufacturer's specified torque.
- Check that the crankshaft turns smoothly. There cannot be any tight spots.
- Follow the same procedure for the other cylinders.

● Putting all the engine fittings back on

- Replace all the components in reverse order of disassembly.
- Replace the bottom plate or sump.
- Refit the cylinder head on a new gasket, following the manufacturer's recommended torque and head bolt-tightening sequence. Refer to the worksheet: 'Replacing the head gasket', page 145.

◆ Testing the engine

Everything has been reassembled. The oil has been replaced. The sea cock is open. The battery is well charged. The fuel lines have been bled. So now you're ready.

But first, it is important to check that the oil pump is working properly. To do this, decompress. If your engine doesn't have a valve-lifter/de-compressor, push on the stop engine button. Activate the starter motor to make the oil rise in the circuit. The engine should turn without any abnormal noise.

Release the stop button or the valve-lifter/de-compressor,

Starting the engine

- Open the throttle slightly. Pre-heat if your engine has glow plugs.
- Activate the starter. After some misfiring, the engine should start. If it does not make sure the fuel system is free of air and bleed the high pressure lines if necessary.
- As soon at the engine starts, make sure that the following function properly:
  - The oil circuit (the oil light goes out or check the pressure).
  - The cooling system (water mixed with exhaust gases comes out of the exhaust).
  - There are no water, oil or fuel leaks.

- Initially it is recommended not to over-rev the engine. It is normal to see a bluish cloud of smoke when first starting. Keep the revs at high idle. After running for a few minutes, progressively lower the revs and listen for abnormal noises.
- Keep a close eye on the water temperature; a sudden rise means that some mechanical part is too tight.
- Check the idle speed.
- Engage the forward gear. Keep the revs at 2000rpm for half an hour. Continue checking. Check the idle and adjust if necessary.
- Is everything OK? Stop the engine.
- Clean the engine bay.

2 Installing the snap clips. When reassembling the piston and rod, make sure they came from the same cylinder. Also, check the rod direction. Don't make a mistake. Generally, on pistons with offset axes, a mark indicates the engine's rotation direction.

3 Generously oil the gudgeon pin and the holes in the pin before assembling.

1 When rebuilding the engine, replace all the old seals with new ones. Manufacturers supply seals individually or in kits for the top and lower parts of the engine.

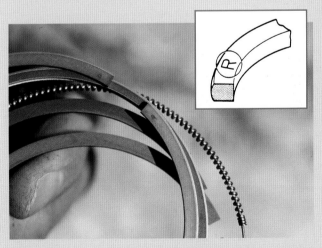

4 Some rings must be mounted in the right direction. A mark is stamped on one of its faces. Always position the brand name or the mark towards the top of the piston.

5 Prior to refitting the rings on the piston, check once more that the grooves are clean. Make sure the ring is free around its entire circumference. If it sticks in its groove it means it hasn't been thoroughly cleaned and must be cleaned again.

6 Proceed with caution when refitting the rings. Start with the lower oil ring and finish with the top compression ring.

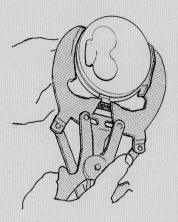

7 The rings are very fragile. If you can, use a ring-removing tool. This way you'll avoid breaking the ring because it has been spread too much, or deformation due to excessive torsion, and damage to the piston.

8 Generously oil the rings while moving them in their grooves.

9 Stagger the ring gaps. Position the gaps in three different directions so they don't face one another. The oil ring gap has to be between 90° and 120° from the piston's axis; 90° when there are three rings and 120° when there are four.

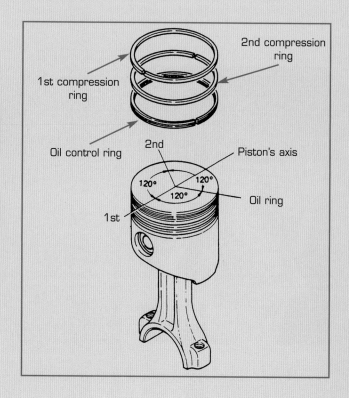

10 When the rings are installed on the piston and the piston rod has been refitted, the rings have to be compressed in the grooves to allow the piston to fit into the cylinder. Be careful not to turn the rings during this operation.

11 Lubricate the shells and journals.

12 Put the piston into the corresponding cylinder and push with a hammer handle. Never force it; guide the rod toward its bearing to avoid scratching its new parts.

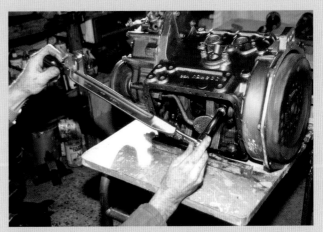

13 Refitting the big end requires tightening them to the manufacturer's specified torque. Use a torque wrench.

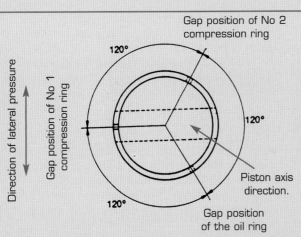

Gap position of No 2 compression ring

120°

Gap position of No 1 compression ring

Direction of lateral pressure

120°

Piston axis direction.

120°

Gap position of the oil ring

The rings have to be placed in such a way that the gaps are 120° apart. Never place a gap on the lateral pressure side.

> ✳ **Pistons with an off-centre axis are placed in such a way that the axis is towards the side with the greatest pressure.**

14 After replacing the oil strainer, put the lower sump back on using a new gasket.

Reconditioning or replacing your engine

Complex
◆ Basic tools

Owing to infrequent use, a sailing yacht's engine will sooner or later begin to 'show its age' and few engines will clock up more than 3000 hours without the occurrence of some mechanical failure. If your engine begins to develop starting or cooling problems you may have to seriously think about reconditioning or even replacing it; but remember, reconditioning might only be viable if your particular engine is still in production and spares are readily available.

Changing your engine for one of a similar size and power should not be a problem but if you would like to replace it with a larger engine having a greater power output, you will need to do some careful research before reaching a decision. (See *How to Install a New Diesel Engine* by Peter Cumberlidge.)

◆ S-Drive transmission

If you have finally decided to replace your boat's engine, the S-Drive might be worth considering as an alternative to your old 'traditional' stern drive? Do your

Before it is ready for bonding in, the bottom face of the S-Drive's fibreglass frame must be carefully shaped to fit the hull.

homework first though as engine price and installation costs vary widely and inevitably, of course, there are advantages and disadvantages attached to both systems.

Mounting essentials

It is important to remember the different steps involved in mounting the propulsion unit onto the engine. Certain procedures must be followed regarding the assembly of the:

◆ Engine
◆ Exhaust system
◆ Water circuit
◆ Fuel circuit

◆ Engine alignment

The alignment of the engine with the drive shaft is a necessary task that must be done with a great deal of care. The final alignment of the couplings is checked with a feeler gauge and must be under five microns for each centimetre of the couplings' diameter.

With the engine out of gear, check the shaft's rotation by hand for any hard spots.

Make sure that the engine is resting squarely and not diagonally on all of its mounts by measuring them to see if they are under equal load.

◆ Exhaust

Install the muffler as close as possible to the engine. Follow the installation measurements. The difference **A** (diagram right) between the upper part of the muffler and the highest point of the lines should not exceed 1.5m. Also the length of the rising lines **B** should not exceed 3m. A longer line requires a larger muffler.

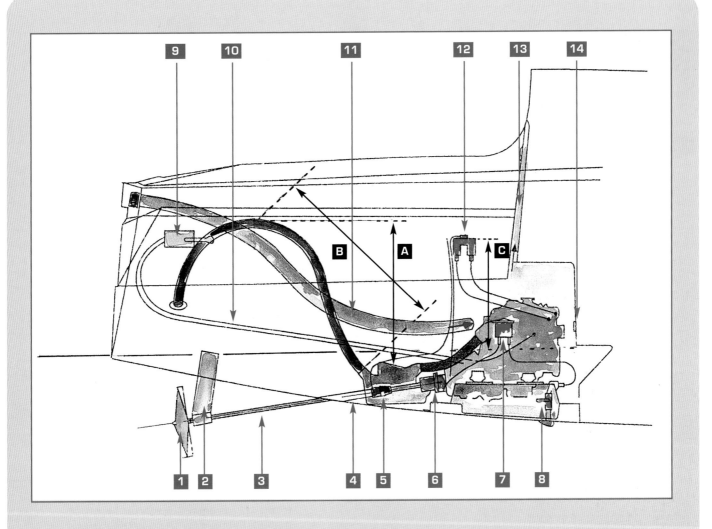

● Water circuit

Check that the water circuit is compatible with the new engine. Use flexible and reinforced suction hoses so that they won't collapse if a vacuum forms. Connections below the waterline need to have double clamps. Avoid sharp bends. To prevent siphoning water out of the engine when it stops, a vent or siphon break must be installed. The distance **C** should be a minimum of 0.4m.

● Fuel circuit

Install the primary filter below the bottom of the tank. The feed and return lines should be of metal (copper or steel). The flexible connection to the engine should meet the ISO 7840 standard.

1 Propeller
2 Strut
3 Propeller shaft
4 Stern tube
5 Stuffing box
6 Coupling
7 Seawater strainer
8 Seacock
9 Throttle control
10 Throttle control cable
11 Fresh air inlet
12 Siphon break
13 Engine room ventilation
14 Fire safety opening

Checking the injection pump timing

Complex

- 1 hour
- Basic tools, injection pump checking tube

A diesel engine is directly dependent on the proper functioning of the fuel injection system. The injectors' condition and the injection timing directly affect the combustion process.

The moment that the jet of fuel is injected into the combustion chamber must be determined with the utmost care. A badly adjusted injection pump lowers performance, increases fuel consumption and causes risk of mechanical damage (backfiring).

When an engine knocks, it means the timing is too far advanced. Black exhaust smoke means the timing is too retarded or that there is a clogged air filter.

This work sheet will help you check the pump timing.

● How to check an in-line pump

The timing adjustment given by the manufacturer determines the position of the crankshaft (marked on the flywheel). As a general rule the mark is before the TDC. Find the timing mark on the engine. The timing mark can be seen through a window cut into the flywheel bell housing and can be found by using a cylinder as a reference – usually the No 1 cylinder, on the flywheel side.

- Disconnect the high pressure lines on the pump.
- Connect an injection pump checking tube.
- With a spanner on the crankshaft nut, turn the engine in its direction of operation.
- Make sure the No 1 cylinder is at the end of compression (this cylinder's valves must not be open; there should be play between the valves and the rockers).
- Slowly turn the engine and align the timing mark with the fixed mark. At that precise moment, the pump for cylinder No 1 must be starting injection.
- The rise of fuel in the high pressure line outlet signals the beginning of injection.

> **! IMPORTANT**
> This type of adjustment requires the utmost care. There must be perfect synchronization between the alignment of the marks and the rise in fuel.

● Adjusting the timing

If the timing mark alignment is wrong:

- Remove the injection pump.

> **! WARNING**
> The procedure for removing the injection pump and the precautions to take differ depending on the manufacturer. You must follow the directions given in the repair manual supplied by the manufacturer.

- Add spacers if the moving mark is too early.
- Remove spacers if the mark is late. Before reinstalling the pump, rotate the camshaft so that the lowest point of the cam(s) are oriented towards the opening where the pump fits.
- Refit the pump.
- Check the timing.
- Re-connect the high pressure lines.
- Bleed the fuel circuit before starting the engine.
- Adjust the idling speed.

If, despite correct timing, the engine still isn't working well, you will have to investigate further by checking the injectors.

1 When removing the pump, first unscrew and remove the high pressure fuel lines. Then unscrew the pump's mounting nuts.
  On Yanmar GM series engines, as shown here, there is no difficulty in removing the pump.

2 To dislodge the pump, tap it with a chisel and hammer on the edge of the flange. Never fit a screwdriver or chisel into the seal seat.

3 Once the pump is loose, push the engine stop lever a few centimetres.

4 Remove the pump; check the number and thickness of the spacers.

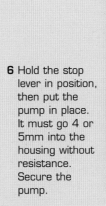

5 Before remounting the pump, set the stop lever so that the slot on the lever lines up with the notch on the injection pump rack.

6 Hold the stop lever in position, then put the pump in place. It must go 4 or 5mm into the housing without resistance. Secure the pump.

7 Find the timing mark on the engine. In the picture (Yanmar engine), a window cut into the flywheel bell housing makes it visible. When the moving mark passes in front of the fixed mark, the injection pump is beginning injection into the cylinder in question.

8 The beginning of injection in a cylinder is indicated by the rise of fuel at the pump's high pressure outlet.

9 Turn the engine very slowly when approaching the injection point so as to get a precise reading.

10 After installing the control tube, turn the engine slowly by hand a few turns to get the fuel to rise in the tube. Adjust the level by using the bleed screw. Injection starts when the fuel starts to rise in the tube.

◆ How to check a distributor-type pump

The timing value defined by the manufacturer determines the position of the crankshaft (marked on the flywheel). As a general rule it is marked before the TDC.

Look for the timing mark on the engine. The mark is visible through a window cut into the flywheel bell housing. Find the timing mark by using a cylinder as a reference – usually the No 1 cylinder, on the flywheel side.

Finding the timing mark on a distributor pump varies according to the type of pump assembly and requires specific tools (pin, comparator) available from the manufacturer. Refer to the maintenance manual.

The operation consists of checking the hydraulic head piston's position at the beginning of injection.

The principle is the same as for the in-line pump. It involves checking the alignment of the injector pump mark with the one on the flywheel.

If, when checking, you notice that the marks are not aligned or that the piston rise doesn't match the timing mark, it is possible to correct that difference:

◆ Loosen the mounting nuts.
◆ Pivot the pump body as necessary.
◆ Re-tighten the pump.
◆ Bleed the fuel lines before starting.
◆ Adjust the idling speed.

If, despite correct timing, the engine still isn't working well, you have to check the injectors.

Checking the timing adjustment on a distributor-type injection pump

**The objective is to synchronise the injection pump movements
with those of the crankshaft and camshaft.**

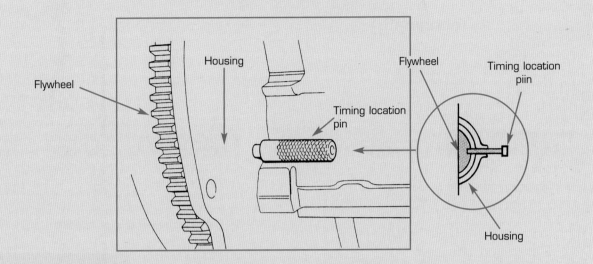

Check the hydraulic head's start of injection
Remove the cover plate to gain access to the markings or
the hydraulic head's bleed screw, depending on the pump
model. Connect the comparator. Find the hydraulic head
piston's BDC. Set the comparator to zero.

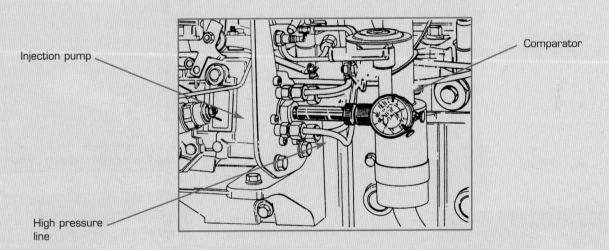

Check the timing mark on the flywheel
Slowly turn the crankshaft in the direction of operation
(clockwise seen from the front) until the pin can fit in the
timing hole, which marks the TDC. Compare the measurement shown with the one given by the manufacturer. If the
measurement is not correct to within 5/100ths of a
millimetre, correct the timing by pivoting the pump body. If
the measurement is too small, turn the pump clockwise
from the rear. Do the opposite if the measurement is too
large. Tighten the mounting nuts. Remove the pin and
make the engine turn about 45° in reverse. Check the
zero on the comparator and check the timing again.

Technical

45 minutes

Spark plug spanner, feeler gauges

Changing the timing belt is part of periodic maintenance. As such, it must be done every 2000 hours or yearly.

Replacing the belt is not particularly difficult but you need to be methodical. First look in the manual for:

◆ The recommended procedure.
◆ The meaning of the marks on the belt and on the pulleys.

Removing the belt

◆ Remove the front protective housing.
◆ Bring the No 1 piston to TDC. The timing mark on the crankshaft pulley is lined up with the one on the housing. (The pulley key is now towards the top).
◆ Remove or release the belt tensioner. When removing it, make absolutely sure that the angle of the two pulleys (crankshaft and camshaft) is not changed to avoid disturbing the timing.
◆ Remove the belt by taking it off of the injection pump pulley.

Timing adjustment; mounting the new belt

Carefully check the alignment of the marks. Fit the new belt, taking into account the direction arrows

printed on it. Start with the camshaft pulley and continue to the crankshaft pulley – not to the ones that are driven (ie water pump, belt tensioner).

Once the belt is in place, set the tension by meticulously following the manufacturers instructions. Make the crankshaft turn two rotations until all the marks are again aligned. There shouldn't be any hard spots. Check and adjust the belt tension if needed.

Replace the protective housing.

2 Remove the protective housing to access the timing belt.

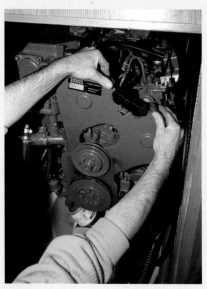

1 Check the camshaft belt every year. It needs to be replaced every 2000 hours or when you notice signs of wear or cracks.

3 Before loosening the belt, line up the timing marks. The synchronous pulleys are fitted on the camshaft 1 and on the injection pump 2. These pulleys are driven by the crankshaft pulley 4 by means of a synchronous belt 3. A free pulley 5 makes the belt run properly. The tensioner 6 allows adjustment of the belt tension.

Diagram 1

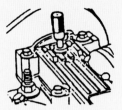

Diagram 2

 If you don't have the manufacturer's manual, check the tension by pushing on the section opposite the one where the tensioner exerts its pressure. It should not flex more than 8mm.

! IMPORTANT
If, when started, the engine 'misfires' or functions abnormally, stop the engine immediately. Check the timing. Without a doubt, one of the two pulleys is off by one notch.

How to change the belt

- Insert the camshaft locking pins (see diagram 1) and the flywheel locking pins (see diagram 2).
- Loosen the camshaft pinion locking screw.
- Insert the injection pump pinion locking pins (see diagram 3).
- Remove the tensioner wheel and the intermediary wheel.
- Remove the belt.
- Install a new belt. Follow the mounting direction arrows.
- Replace the intermediary wheel and the tensioner wheel.
- Remove the pump locking pins.
- Set the belt tension (see diagram 4).
- Tighten the camshaft pulley nut to the specified torque.
- Remove the locking pins on the camshaft and the flywheel. Check the belt tension.
- Check the injection pump timing.

Diagram 4

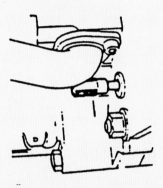
Tension checking tool

Diagram 3

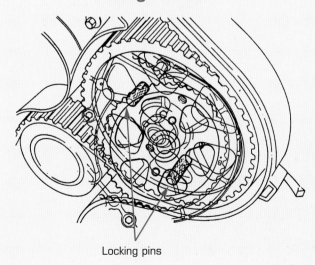

Locking pins

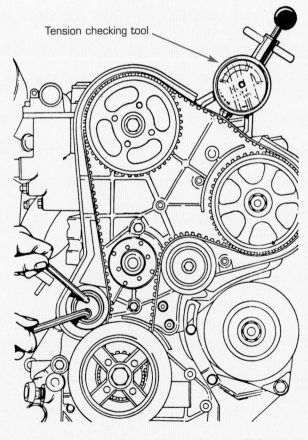

Adjust the tension by pivoting the tension wheel. Tighten.

Adjusting the injection pump timing

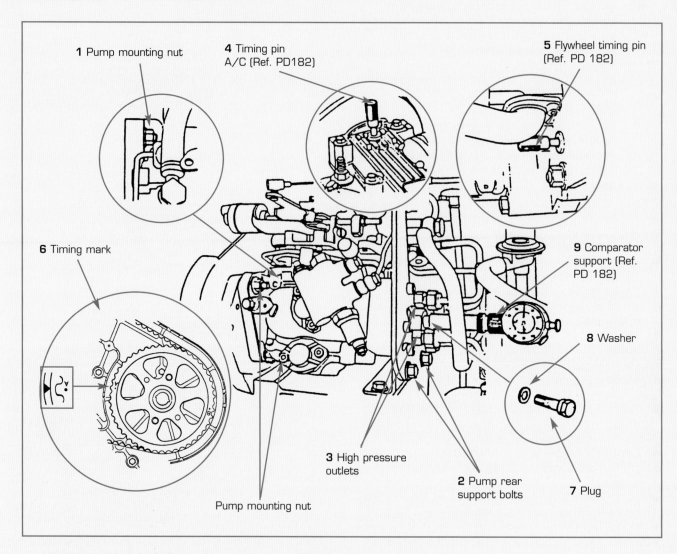

1 Pump mounting nut

4 Timing pin A/C (Ref. PD182)

5 Flywheel timing pin (Ref. PD 182)

6 Timing mark

9 Comparator support (Ref. PD 182)

8 Washer

3 High pressure outlets

2 Pump rear support bolts

7 Plug

Pump mounting nut

Timing method

1 Position timing pins (4) and (5). The No 1 piston is then at TDC, compression stroke. In case of difficulty inserting the pin in the flywheel, unscrew the camshaft pinion bolt and turn the crankshaft.
   Tighten the camshaft pinion bolt.

2 Make sure the pump mark is correct (See 6).

3 Remove the plug (7). Put the support (9) and comparator in place. Adjust it to 2mm.

4 Remove the pins (4) and (5) and turn the crankshaft anti-clockwise until the comparator indicates the pump piston's BDC. Set the comparator at zero.

5 Turn the crankshaft clockwise until it is possible to engage pins (4) and (5) in their places.
   The comparator should now show the pump timing value (eg 1.37mm Prima; 1.17mm Prima Turbo).

6 If the value shown is incorrect:
   ◆ Disconnect the high pressure outlets (3).
   ◆ Loosen the rear support bolts (2).
   ◆ Loosen the pump mounting nuts (1).
   If the piston reading is **too low**, turn the pump toward the engine.
   If the piston reading is **too high**, turn the pump away from the engine.

7 Tighten the pump mounting nuts.

8 Remove the pins (4) and (5) and turn the crankshaft about $1/4$ turn anti-clockwise.

9 Turn it back clockwise until the pins fit in their holes.

10 See if the piston reading is correct.

11 Correct and check again if needed. Tighten the rear pump mounting bolts, the pump mounting nuts (1) and the high pressure outlet connections.

Checking the injectors

Simple

- 15 minutes
- Basic tools

The injection pump is in perfect condition; the timing has been checked and the system has been bled. But the engine still emits a bluish smoke, especially when at high revs. It is possible that one or more injectors are faulty.

A quick test allows you to check the combustion, and therefore the injector function.

 WHAT THE TEST TELLS YOU
- ◆ Each injector that is loosened should cause a corresponding drop in revs.
- ◆ When loosening an injector, if the engine revs don't change, the injector is faulty.

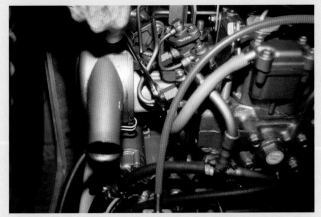

1 Unscrew the high pressure line one turn and listen for a drop in revs.

2 A white rag placed under the line allows you to see the amount of fuel injected and prevents diesel from spilling onto the hot cylinder head.

3 When you retighten the line, the engine should return to its initial revs.

4 Repeat the operation on the other cylinder(s).

Removing and checking an injector

Complex

- 1 hour
- Basic tools

The injector atomises the fuel received from the injection pump and injects it into the pre-combustion chamber or above the piston crown.

Engine function is extremely sensitive to injection pressure, atomisation quality and the injector's seal.

Bad atomisation or a leak invariably causes lower engine performance. The engine doesn't burn all the fuel, looses power, knocks a lot and the idle is erratic.

The principal parts of the injector are the nozzle and the needle. These parts are exposed to high combustion gas pressures and high temperatures, they need to be checked every 1000 hours.

This job has to be done with the engine cold.

> **!** **IMPORTANT**
> Removing and checking the injectors requires absolute cleanliness.

Removing the injector should be fairly easy. However if you have any difficulty removing it, it is possible that it has seized in the socket. Spray it with some penetrating oil and pry it loose by using a lever between the injector holder and the cylinder head. It is best to use an injector remover for this job. The tool, which screws in place of the fuel line, can be home-made.

You can check the injectors by reconnecting them to the injection pump's high pressure lines without putting them back in the sockets. Be careful not to bend the lines. Activate the decompression lever and crank the engine with the starter motor. The jet has to be regular, uniform and symmetrical. If it is a multiple-hole injector make sure all the holes are spraying.

1 Clean around the injectors then begin by removing the leak-off pipes.

2 Loosen the high pressure lines.

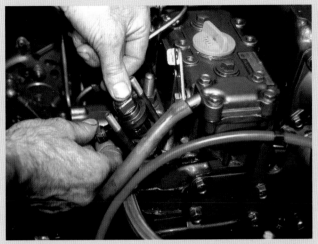

3 Loosen and remove the injector holding plate.

4 Move the high pressure line; remove the injector.

5 To remove the pre-combustion chambers (Yanmar), activate the starter motor. The air compressed by the piston will push out the pre-combustion chambers.

6 Plug the injector opening on the head so no grit or foreign bodies can get into the engine.

7 At this stage, it is easy to see the injector function by re-mounting it on the high pressure line. With the compression chambers removed, push the decompression lever and crank the engine with the starter motor. The injector should spray a fine mist of fuel.

Reconditioning the injectors

Technical

- 1 hour per injector
- Basic tools, injector calibrating pump, torque wrench

This job should be done by an injection specialist. You can only do it yourself if you have an injector calibrating pump.

Disassembly

Unscrew the nut holding the injector nozzle to release the spring tension. Separate all the parts and place them in a container filled with well-filtered fuel.

> ### IMPORTANT
> The dismantled parts must be cleaned with new, filtered diesel fuel. Carbon deposits must be removed from any dirty parts.
> Each injector assembly must be placed in a separate container, so they cannot be mixed up.

Inspection

Check the needle. No trace of pitting or scratches are allowed on the needle. Check that the nipple or the needle point is in perfect condition.

Check how the needle slides in its seat by tilting it 45°; the needle should fall back under its own weight. If the needle slides irregularly or sticks, replace the injector.

> ### IMPORTANT
> Never touch the needle's body. Handle it with the little rod on the spring side. Before engaging the needle in the injector, dip both parts in clean diesel.

Check the calibration

With the nozzle holder connected to the calibrating pump, slowly activate the pump lever. Read the opening pressure. The calibration is made by adjusting the spring pressure with spacing shims. If the pressure is lower than prescribed, add more spacers under the spring; if it is too high, remove some spacers.

Check for leaks

Check that the injector doesn't drip or dribble. To do that, raise the pressure until it is 20 bars below the injection calibration. If a droplet forms, the injector leaks. Replace the injector.

Check the spray

Close the valve on the gauge. Check the spray quality by rapidly activating the pump.

To reassemble the injector

- Clean the injector socket.
- Change the copper washers.
- Lightly oil the injector holder.
- Refit the injector in the cylinder head. It should fit without any hard spots and without forcing. If not, carefully clean the injector socket.
- Fit the high pressure lines on the injector holder without tightening.
- Tighten the injector holder to the manufacturer's recommended torque.
- If there is a holding plate, use a torque wrench to tighten the fastening bolts evenly and alternately.
- Tighten the high pressure lines.
- Reinstall the leak-off pipes.
- Finally bleed the system as previously explained.

> If there is a leak in the high pressure lines, don't tighten further. Loosen and retighten.

1 To disassemble an injector, it must be held it in a vice fitted with soft jaws.

2 Separate the different parts and place in a clean container filled with diesel.

3 Adding or removing small shims allows the injection pressure to be adjusted.

4 After washing the injector nozzle with diesel, check that the needle slides easily, it should drop under its own weight. If the needle snags or slides irregularly, change the injector nozzle.

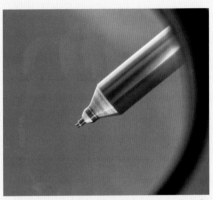

5 A magnifying glass is used to check the needle to detect the slightest scratch or seizing.

6 Install the injector on the test pump. Slowly raise the pressure and note when the injection starts. If the pressure is lower than that prescribed by the manufacturer, add wedges onto the spring. If the pressure is higher, remove the wedges. The injection pressure will vary 10 bars/cm for each 0.1mm of shim.

7 Check the spray quality by rapidly causing injection. The spray should be regular, uniform and symmetrical. In the case of a multi-hole injector, make sure that all of the holes are clear.

8 After the spray test, wipe the nozzle to see if it leaks. Keep the pressure 10 bars below triggering pressure. If you see a droplet forming, change the injector.

Replacing the starter motor

Simple
- 30 minutes
- Basic tools

Now that crank handles have almost disappeared from engines, keeping your starter motor in perfect condition is essential.

Just like all mechanical and electrical parts, the starter motor can break down. Any reconditioning – either complete or partial – will require its removal from the engine.

Removing the starter motor

- Switch off the starter battery.
- On a piece of paper, draw the connection positions, the wire colours and their respective positions. As you disconnect the wires, check the crimping as well as the condition of the terminals to see if they are clean.
- Unfasten the starter's mounting bolts. The motor being heavy, it is best to loosen all the bolts without removing them, while holding the starter steady.
- Remove the starter.

Refitting the reconditioned or new starter motor

- Put the starter back in place.
- Screw in the mounting bolts while supporting the starter.
- Connect the positive cable (heavy gauge wire), then the ignition switch wire. Switch off the starter battery.
- The engine has to be run at a speed above 150rpm.

Starter motor wiring circuit

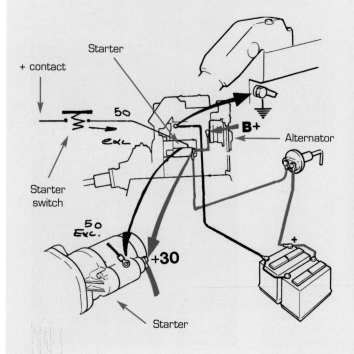

1 To avoid an electrical short, before doing any work on the engine, especially on the electrical circuit, remember to open the circuit breakers.

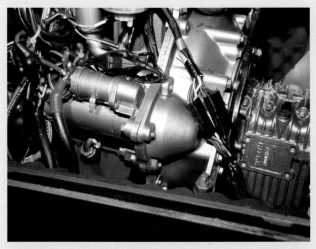

2 As a precaution, before disconnecting the wires, draw the connection positions, the wire colours and their respective positions on a piece of paper.

3 Disconnect each wire and clean the terminals and also check the state of the crimping.

4 Electrical tape is an efficient way to mark them.

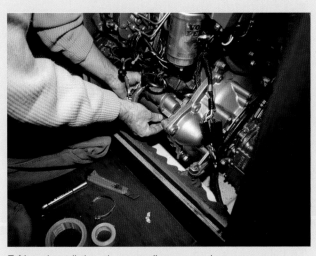

5 Now that all the wires are disconnected, you can remove the starter's mounting screws.

6 Start by removing the bolts that are the hardest to reach; end with the easiest.

7 Remove the starter, but be careful – it's a very heavy part.

Replacing the alternator

Simple

- 45 minutes
- Basic tools

On a boat, as with a car, the battery only stores a small amount of power, which is used to start the engine. An alternator is needed to recharge the battery. This part has excellent reliability and requires minimum maintenance, but if you need to inspect it, it has to be removed.

● Removing the alternator

- ◆ Open the circuit breaker. On a piece of paper draw the connection positions, the wire colours and their respective positions.
- ◆ Disconnect each wire while checking the condition of its terminals – the crimping and whether they are clean.
- ◆ Loosen the pivoting axis and the adjusting bolt or nut without completely unscrewing them.
- ◆ Loosen, then remove the belt.
- ◆ Remove the adjusting and mounting bolts.
- ◆ Remove the alternator.

● Refitting the alternator

- ◆ Put the alternator in place and insert the pivoting bolt and the adjusting bolt, without tightening them.
- ◆ Check the condition of the belt. Do not hesitate to replace it if necessary. Put the belt in place and push the alternator with a lever, then tighten the adjusting bolt. Check the tension by pushing hard in the middle of its greatest length (for information, refer to the work task 'Adjusting the belt tension').
- ◆ Tighten the pivoting axis.
- ◆ Re-connect the wires following the information recorded when you dismantled it.
- ◆ Check the alternator's charge; the indicator light on the control panel should go off. If in doubt, check the charging voltage.

Charging circuit wiring diagram

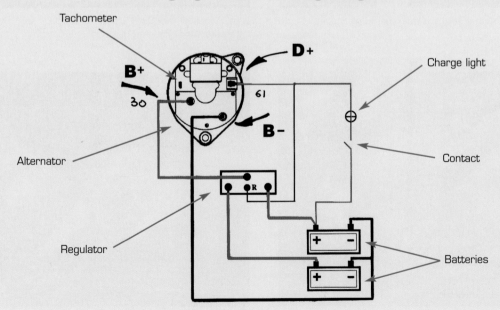

1 Note the alternator's position and its accessibility. On a paper, draw the position of the connections, the wire colours and their respective positions.

2 Disconnect each wire. Put the nuts back on their respective screws to avoid losing them.

3 Loosen the pivoting axis and the adjustment bolt/nut.

4 Then push the alternator to remove the belt.

5 Remove the adjusting bolt then slide out the pivoting axis while holding the alternator.

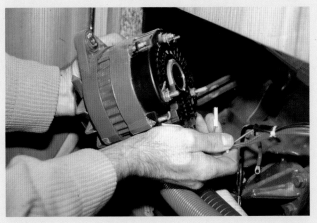

6 Remove the alternator.

Checking the glow plugs

Complex

- 45 minutes
- Basic tools, test lamp, multimeter, battery cable

Starting an indirect injection diesel engine differs a bit from its directly injected counterpart. Indirectly injected engines need to heat the pre-combustion chambers to be able to start because the temperature at the end of the compression is not high enough to cause auto-ignition of the fuel.

If your engine is difficult to start when cold, it is possible that the pre-heating system is defective.

It is easy to check whether one or more glow plugs are faulty.

Testing a glow plug

- Disconnect the bar connecting the glow plugs.
- Connect a test lamp in series to each glow plug, one after the other.
- If the light comes on, the glow plug is good. If it stays off, the plug has burnt out.
- Replace the plug.

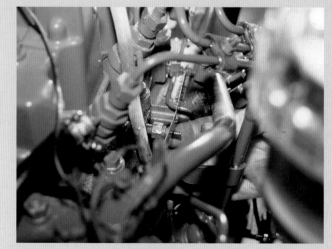

2 Be careful not to lose the nuts and washers.

1 The glow plugs are often the cause of poor engine starting. To check them, first remove the connecting bar.

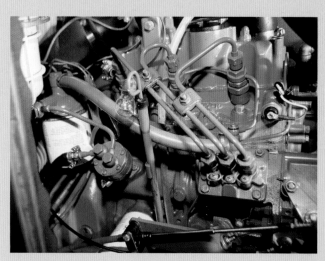

3 Connect a test lamp in series between the live wire and the glow plug.

4 Activate pre-heating. You must get help for this operation so the glow plug switch can be pushed while you look at the test lamp.

5 If the test lamp comes on, the plug is good. If it stays off, the plug is dead and you need to change it.

◆ Visual test of glow plug

It is possible to see how well a glow plug is working. To do this: remove the plug. Feed it with approximately 10 volts of power. After a few seconds the plug should heat up, then glow. To avoid damaging the plug, cut the power after 10 seconds.

> **! IMPORTANT**
> **Remember to coat the plug thread with graphite grease to make it easy to remove later.**

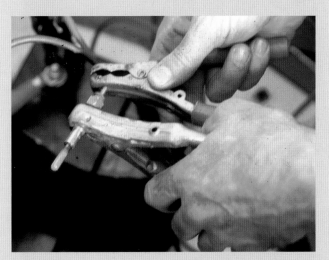

6 Unscrew and remove the plug.

7 Feed the plug with about 10 volts of power with a battery jump cable. A lightly discharged battery connected in series can do the job.

8 After a few seconds, the plug should heat up, then glow on most of its length. To avoid damaging the plug, cut the power after 10–15 seconds. If the plug doesn't work, it is best to change them all to keep the pre-heating system in balance.

Checking the charging system

Simple

- 15 minutes
- Ammeter with clips, multimeter

The alternator and its regulator will need to be checked to address any of the following symptoms:

- Abnormal charge light function
- Undercharged battery, shown by slow cranking
- Overcharged battery

When the charge light stays on it means the charge is insufficient. Undercharging may be caused by a faulty battery, a loose belt, bad connections, a badly set regulator or a faulty alternator. Overcharging is caused by a badly adjusted regulator. It causes excessive battery water consumption.

How to do it

- Get a voltmeter, preferably digital.
- Check the belt condition and tension.
- Check the tightness of the connections on the alternator, the starter and the battery.

> **! IMPORTANT**
> **Never disconnect the alternator when the engine is running.**

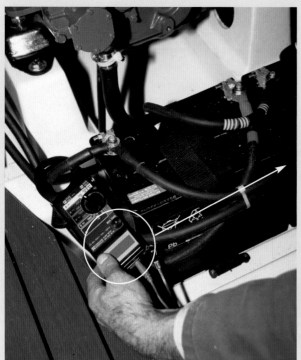

`000 A`

Checking for stray current
This test, which is done with an ammeter, shows any current leakage.

Test conditions
Engine stopped, contact off and nothing turned on. Attach the ammeter clips on the positive or negative battery cable. (Check the direction of the crocodile clips.)

Reading
There should be no current leak.

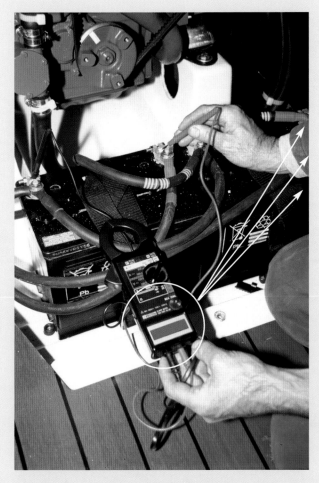

16.5 v	Change the regulator.
13.7 v	Charging is normal.
12 v	The circuit is not charging; check the alternator.

Checking the charge voltage

This test is done with a voltmeter connected in parallel to the battery.

How to do it

◆ Check the battery charge.
◆ Start the engine.
◆ Slowly rev it up and note the change in voltage.

What it means

◆ If the voltage stays the same despite the change in revs, the alternator is not charging.
◆ If the voltage is above 15 volts, the regulator is not working.
◆ If the voltage stabilises at 13.5–14 volts, the charging system is doing its job.

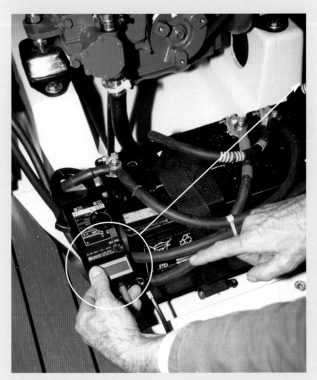

| 25 A |

Test the output

This test is done with an ammeter with crocodile clips. It shows the alternator output or how many amps it is putting out.

How to do it

◆ Put the ammeter clip on the positive cable or negative battery cable.
◆ Check the direction of the clips.
◆ Start the engine.
◆ Note the output at idle.
◆ Slowly rev up and note the reading.
◆ Turn on every appliance and light and repeat the operation.
◆ Note the reading.

What it means

Compare the readings (number of amps) with the manufacturer's specifications.

Technical

- 2 hours
- Basic tools, multimeter

We never think about the starter motor whilst it is working. However, checking it regularly should be part of your maintenance routine. It also must be checked whenever one of the following symptoms appears.

Symptoms

- ◆ The starter doesn't crank.
- ◆ The starter turns the engine too slowly to start it.
- ◆ The starter turns but not the engine.

Test conditions

- ◆ A fully charged battery.
- ◆ Start switch in perfect condition.

Exploded view of a starter motor

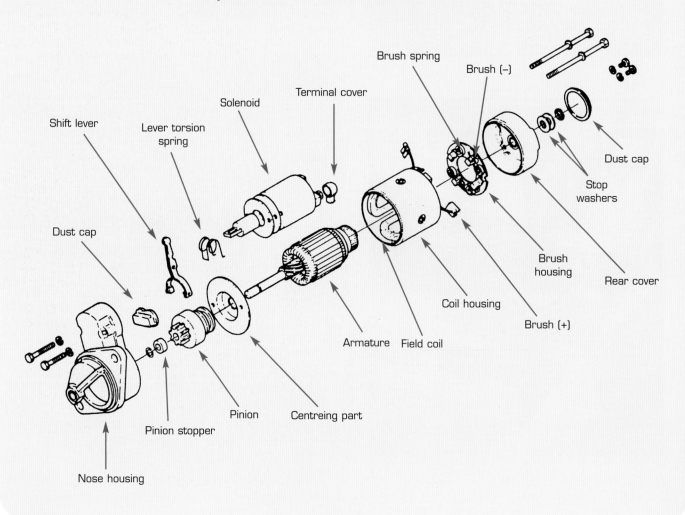

Disassembling the starter motor

- Clean the starter exterior with a brush and degreaser.
- Disconnect the coil feed cables.
- Remove the armature screw and locking system.
- Remove the screws holding the end frame and remove it.
- Remove the coil holder.
- Chase out the shift fork pin.
- Unscrew and remove the solenoid screws.
- Take out the armature.
- Make the pinion's stopper slide below the snap ring.
- Remove the snap ring.
- Remove the pinion.

1 Hold the starter in a vice and remove the dust cap.

2 Remove the snap ring and the stop washer. Be careful not to lose the washer and the spacer. Then remove the two screws holding the rear cover and those holding the brush holder.

3 Remove the coil wires then unscrew the two holding the solenoid.

4 Disengage the negative brush. Take the positive brush out of the brush holder.

5 Remove the coil holder and the armature assembly.

6 Slide the snap ring out of the armature shaft with a small screw driver.

8 Remove the starter pinion from the armature shaft

7 Punch out the pinion stopper behind the snap ring situated at the end of the armature assembly, drive side. Use a tube with an internal diameter slightly larger than the shaft to do this. Here a ratchet wrench extension is being used.

9 Clean all the parts.

Visual and mechanical inspections

◆ Examine the case and field coils. Look for traces of friction, indicating an out-of-centre armature.
◆ Check if the commutator is out of round (0.05mm maximum). If necessary, machine lathe the surface and clean the slots with a ground hacksaw blade.
◆ Check that the Bendix clutch spins freely in one direction and holds tight in the other. If the Bendix spins freely both ways, replace it. Visually check the Bendix pinion. Replace it if it looks worn.
◆ Placing the armature shaft in both end frames, check the play in the bushings. There shouldn't be a lot of play.
◆ Change the bushings or bearings if needed.
◆ Check the brushes' condition, wear and how they slide. Don't hesitate to change them if they are worn because they are very important to the starter's performance. If soldered on, cut them near the connecting bar or close to the rear, remove all traces of varnish and solder on new ones.
◆ Visually check the solder condition, the commutator bars, the coils.

11 Check the insulation between the commutator bars. Lower the insulation level with a hacksaw blade when the mica is too close to the surface.

10 Check the commutator's surface; clean it with fine sand-paper. If it is damaged it has to be machined to the manufacturer's specifications.

Checking the insulation cuts

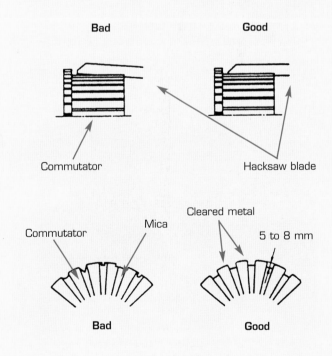

Bad   Good

Commutator   Hacksaw blade

Cleared metal

Commutator   Mica   5 to 8 mm

Bad   Good

12 Replace the Bendix clutch if the pinion looks worn. Check the clutch mechanism. It has to spin freely one way and lock the other. If it spins freely both ways, the clutch mechanism is faulty and has to be replaced.

13 Check the front and rear bushes for ovalisation.

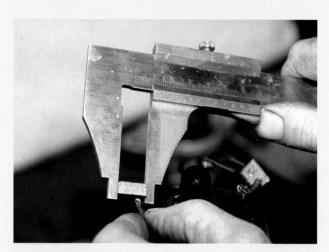

14 Check the measurement of the brushes to know how much useful life they have left. Replace them if they are worn beyond the allowed limit. If they need replacing, cut the brushes near the connecting bar for the positive brush and near the brush holder for the negative; remove all the varnish and use a soldering iron to fit new ones.

Electrical inspections

Check:

- ◆ The commutator's insulation.
- ◆ The insulation and continuity of the field windings.
- ◆ The insulation on the positive brush holder.
- ◆ The continuity on the negative brush holder.
- ◆ The continuity of the solenoid's trigger coil.
- ◆ The continuity of the solenoid's starting coil.

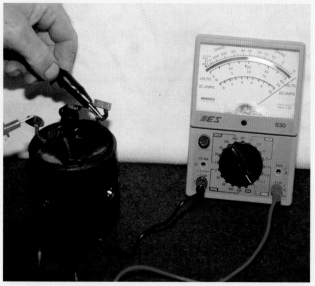

15 Test the continuity of the field winding. If the resistance is good, the coil is good. Also check its insulation by checking the continuity between the end of the coil and the case. If there is continuity, the coil is no longer insulated and it needs replacing.

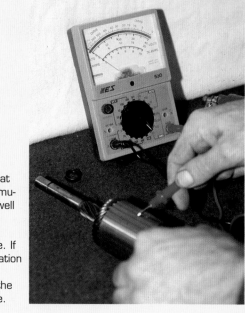

16 Check that the commutator is well insulated from the armature. If the insulation is bad, replace the armature.

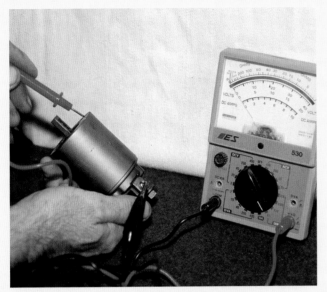

17 When testing a solenoid, check the continuity of the excitation coil and the starter coil. If the continuity is good, the excitation and starter coils are good.

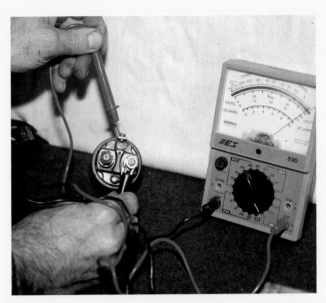

18 The solenoid excitation coil is checked between the excitation coil terminal and the exit terminal on the induction coil side. The starting coil is checked between the excitation coil and the ground.

Reassembly

After you have replaced the faulty parts, reassemble the starter motor in reverse order to that in which it was dismantled. Make sure that the insulating washers are well positioned and that the brushes are held inside the holders when fitting the armature. Take care to lightly oil the shaft on the Bendix side. Make sure to correctly orient the shift lever, solenoid nose housing, case, and end frame. Centreing lugs allow the parts to be centred correctly. Once reassembled, spin the rotor by hand. Then before remounting the motor, test it on a bench.

19 When reassembling the Bendix onto the armature, lightly lubricate the shaft, fit in the Bendix stopper and slide the snap clip into its groove.

20 Force the stopper in over the clip with one or two pairs of pliers.

21 Reassemble the starter in the reverse order of dismantling. Lightly grease the front and rear bushes and remember the washer and the spacer. Replace the snap ring with a pair of pliers.

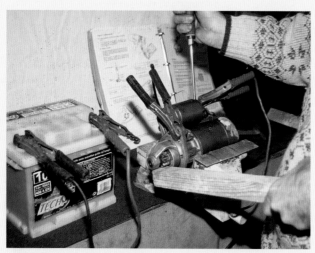

22 In a vice, check the starter for correct rotation and torque. This is checked by braking the starter with a piece of wood. If the torque is strong and the revs high, the starter is in good condition. Slow rotations and weak torque means a faulty starter motor.

23 To start the motor, use a screw driver to connect the excitation terminal with the starter feed cable.

● Bench testing

◆ Check the pinion rotation by hand. It should spin only one way.
◆ Hold the starter in a vice.
◆ Connect the starter as indicated in the drawing below.
◆ Feed power to the solenoid. The starter should spin immediately at its maximum revs. Check that the pinion slides instantaneously on the helical grooves and returns all the way when the starter is stopped.

Starter motor test

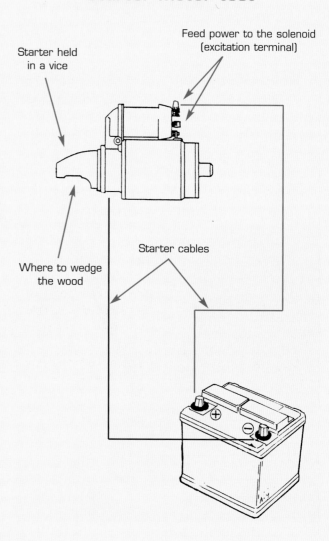

Starter held in a vice

Feed power to the solenoid (excitation terminal)

Starter cables

Where to wedge the wood

Troubleshooting table for starter motor problems

PROBLEM	CAUSE	SOLUTION
The pinion doesn't go forward when the contact is on		
Circuit	Battery or switch terminals loose or removed	Repair or re-solder
Switch	Faulty connections	Replace the switch
Starter	The ramps are damaged, the pinion can't move	Replace
Magnetic relay	The plunger doesn't slide or the coils are bad	Repair or replace
The pinion is engaged and the starter turns, but the engine is not cranked		
Starter motor	Faulty pinion clutch	Replace
The starter spins at full speed before the pinion is engaged on the flywheel		
Starter	Deformed lever torsion spring	Replace
The pinion engages on the flywheel but the starter doesn't crank		
Circuit	Battery cables, solenoid, faulty grounds or loose battery terminals	Repair, tighten or replace wires
Starter	Bad pinion flywheel engagement Bad starter assembly Worn brushes or faulty springs Dirty commutator Faulty induction or field coil Loose field coil or brush terminals	Replace Check the assembly Replace Clean Check or replace Tighten
Magnetic relay	Pitted or defective connections	Replace
The starter doesn't stop after the engine runs and the contact is cut		
Switch	Faulty switch	Replace
Magnetic relay	Faulty magnetic relay	Replace

Checking the alternator

Complex

⬡ 3 hours

◆ Basic tools, soldering iron, multimeter, vice

In general, modern alternators are extremely reliable and require almost no maintenance. Even so, with time the alternator output can begin to decrease. So you need to check for such faults as worn brushes and bearings and, if necessary replace the alternator.

First, look for the fault before deciding to remove the alternator and recondition it.

If the regulator's voltage output is too weak, first check the battery and its connections, before starting on the alternator.

If the alternator is noisy, make sure the noise really comes from it. Put a stick or a plastic hose on the end plates and listen for the precise source of the noise. If the alternator doesn't charge, first check the brushes. If it charges too much, change the regulator.

Exploded view of an alternator

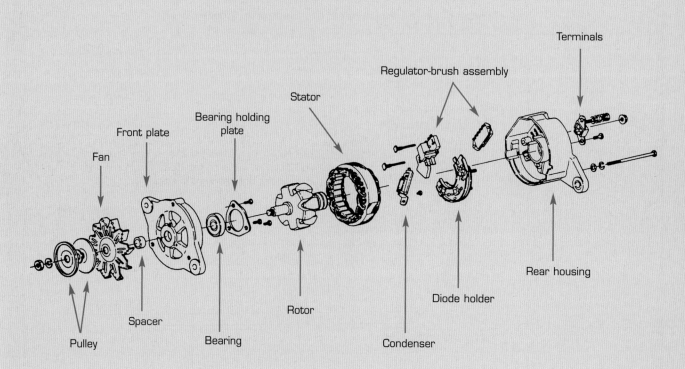

◆ Dismantling the alternator

◆ Before taking the alternator apart, make some alignment marks on the case and end plates. These marks will help you to reassemble the parts correctly.
◆ Remove the drive pulley, held by a central nut.
◆ Remove the key.
◆ Remove the brush holder, or when applicable, the regulator brush holder.
◆ Unscrew the retaining screws and remove the rear plate.

At this stage on some models you have to unsolder the diodes.

! PRECAUTION
Find the wires from the stator to the diodes and unsolder them, but first hold the diode wire with pliers so it doesn't heat up.

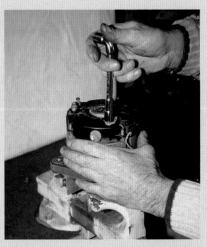

1 First, remove the protective cover, then the brush-regulator holder.

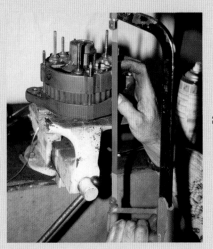

2 Before dismantling the alternator, make some alignment marks for the end plates. In the photo, a small cut made with a hacksaw will help to reassemble correctly.

◆ Changing the bearings

Remove the snap ring with pliers, then chase the bearing from the casing. Extract the bearing from the rotor shaft with a three-jawed puller.

3 To remove the pulley, immobilise its spindle with an Allen key in a vice or hold the pulley with an old belt held in a vice. Unscrew the nut.

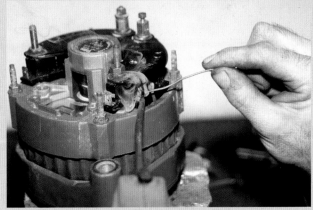

4 On some models, the stator has to be unsoldered to remove the diode holder. Be careful not to overheat the diodes. Put a pliers on the diode wire to absorb the heat when you're unsoldering the stator.

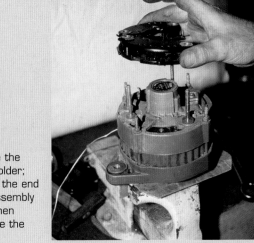

5 Remove the diode holder; remove the end plate assembly bolts, then separate the plates.

◆ Visual and mechanical tests

- ◆ Check the bearings by turning them by hand. There must not be any hard spots or play.
- ◆ Check the condition of the rotor commutator. Polish it with fine sandpaper. Note that it is not possible to machine it. If the ferrules are worn, replace the rotor. Once reassembled, check that the rotor spins freely by hand.

- ◆ Measure the length of the brushes and check that they slide properly in their holder. Make sure they stick out at least 5mm from the holder.
- ◆ Check the springs.
- ◆ Inspect the coils: their smell or colour may give you some clues.

6 Clean the ferrules with fine sandpaper. If the ferrules are worn, replace the rotor.

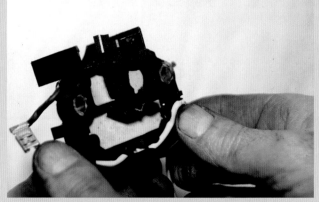

7 Measure the length of the brushes. On some models, the brush holder is integrated into the regulator.

◆ Electrical tests

The electrical tests are made with an ohmmeter.

Rotor: check the resistance (between 2 and 10 ohms) and the rotor's insulation.
Stator: check the resistance and insulation of each stator coil.

Diode: check the diodes with an ohmmeter. The current should only flow one way.
Brush holder: Check the insulation of the positive brush holder relative to the ground and the continuity on the negative brush.

8 Check the diode board. Place one ohmmeter tip on the diode support and the other on each of the outlets. Reverse the ohmmeter tips and repeat the operation. The current must only flow one way.

Electrical diagram of a diode board

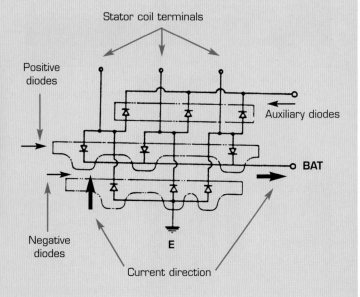

Stator coil terminals

Positive diodes

Auxiliary diodes

BAT

Negative diodes

E

Current direction

Test the stator

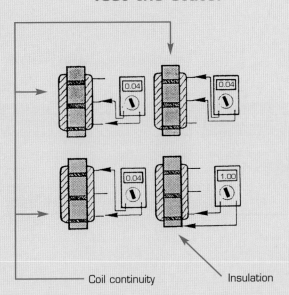

Coil continuity   Insulation

9   Testing the stator. Check each coil individually. The resistance should be the same for each coil (0.1 to 0.7 ohm). Check the insulation of the coils from the casing. (The resistance should read infinity.)

Test the rotor

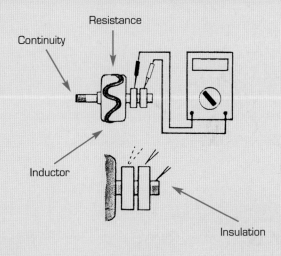

Continuity   Resistance

Inductor

Insulation

10   Test the rotor. Check the stator resistance (2 to 10 ohm) by placing the ohmmeter tips on each ferrule. Check the ferrule's insulation. (The resistance should read infinity.)

● Reassembly

After having replaced the faulty parts, reassemble the alternator in the reverse order to dismantling. Be careful not to overheat the various parts when soldering the stator connectors. Make sure the end plates and the stator are aligned. Remember to grease the bearings.

Progressively tighten the assembly bolts. Once the alternator is reassembled, check by hand that the rotor spins freely.

The alternator is ready to be remounted on the engine.

Troubleshooting table for alternator problems

PROBLEMS		CAUSES	SOLUTIONS
Not charging	Circuit	Wires broken, short-circuited or disconnected	Repair or replace
	Alternator	Coil damaged, short-circuited, or grounded. Faulty terminal insulation	Repair or replace
	Regulator	Faulty regulator transistor	Adjust or replace
Insufficient battery charge, rapid	Circuit	Loose or broken wires, incorrect wire section or length	Repair or replace
	Alternator	Rotor coil short-circuited.	Replace
		Stator coil short-circuited; broken stator line.	Replace
		A dirty brush ring	Clean and polish
		Loose belt	Tighten
		Bad brush connections	Repair
		Faulty diode.	Replace
Overcharged battery	Battery	Electrolyte level low or incorrect density	Add distilled water Check density Replace battery
	Regulator	Faulty transistor	Replace the regulator
Unstable charging current	Circuit	Broken wire making intermittent contact	Replace or replace
	Alternator	Faulty insulation	Replace
		Faulty brush spring	Replace
		Dirty brush ring	Clean and polish
		Damaged coil	Repair or replace

Checking and measuring stray current on the engine

Simple

- 30 minutes
- Basic tools, digital multimeter

If your installation is well designed, it shouldn't have any stray current. The most visible indicator of stray current is the fast decomposition of the protective anodes. Any suspected current loss will necessitate checking the entire electrical circuit.

 IMPORTANT
The engine must be electrically insulated

Stray current can be detected with a multimeter or simply with a test lamp.

Using a multimeter

Test conditions: Contact in 'On' position and no other battery drain. Connect the multimeter in series to the battery's positive terminal.

Reading:

- A few milliamps: all is normal.
- Several dozens milliamps: the system is faulty.
- One amp or more: there is a short.

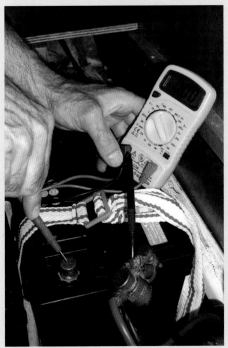

The use of a digital multimeter makes sure you get a precise reading showing the decimals and the polarity.

Using a test lamp

Test conditions: Contact in 'On' position and no other battery drain. Put the test lamp between the positive battery terminal and the positive lug.

Reading: Even a slight warming of the filament means there is stray current. This is grounds for further investigation.

Made up of two wires soldered on to a bulb, the test lamp is the simplest instrument to check for stray current.

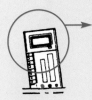

Connect the multimeter in series
With the multimeter in amp position, set mA.

- A few milliamps: all is normal.
- Several dozen milliamps: the system is faulty.
- One extra amp: there is a short.

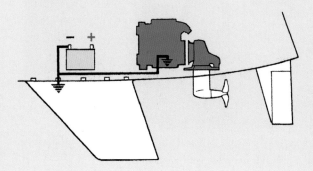

Checking and measuring stray current on an S-Drive transmission

Simple

- 30 minutes
- Basic tools, digital multimeter

> **! IMPORTANT**
> The flywheel bell housing and the lower leg should never be connected to the engine ground

With the multimeter on the ohmmeter position, check the lower leg's insulation from:

- The engine.
- The hull (when metallic)

No insulation defect is acceptable.

If the drive has an insulation fault, its cause must be found:

- Flywheel housing.
- Gearbox controls.

Also check that no wires are screwed onto the lower leg housing.

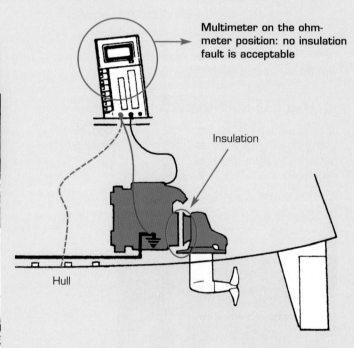

Multimeter on the ohmmeter position: no insulation fault is acceptable

Insulation

Hull

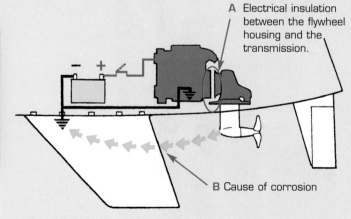

A Electrical insulation between the flywheel housing and the transmission.

B Cause of corrosion

If you note an insulation fault, you must find the source to avoid significant damage from galvanic corrosion.

An insulation fault **A** closes the circuit and causes corrosion **B**.

Changing the protective anodes on an S-Drive

Simple

◆ 1 hour
◆ Basic tools

To replace the anodes on an S-Drive, the propeller must be removed. The anode should be replaced each sailing season or when it is 50 per cent worn.

● How to do it

◆ Unscrew the locking screws on the end nut holding the propeller.
◆ Unscrew the end nut with the help of an Allen key.
◆ Remove the propeller.
◆ Carefully clean the shaft and the propeller hub.
◆ Unscrew the fastening bolts and remove the anode.

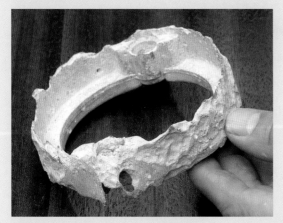

Definitely time this anode was changed!

Block the propeller with a big piece of wood. Then undo the fitted nut with the help of an Allen key.

◆ Put on a new anode and tighten the fastening bolts. Make sure there is good mechanical contact.
◆ Lubricate the shaft and the propeller hub. (Suitable grease: eg VOLVO 82825061)
◆ Mount the propeller, tighten the holding nut and the locking screws.

● A folding propeller

◆ Install the propeller hub 3 on the shaft after you have greased and tightened the counter nut 6. Use a 24mm socket to do this.
◆ Grease the blade pins 8 and teeth 2.
◆ Mount a blade on the hub and push in the pin until the groove 9 is exactly at the centre of the locking screw 7.
◆ Tighten the locking screw with a 4mm Allen key.
◆ Do the same to mount the other blade.
◆ Make sure the blades are at the same angle to the shaft and that they open freely.

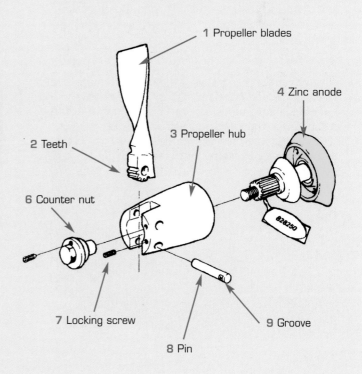

1 Propeller blades
4 Zinc anode
3 Propeller hub
2 Teeth
6 Counter nut
7 Locking screw
9 Groove
8 Pin

Complex

- 1 hour
- Basic tools, compression gauge

This object of this test, which is to check the air tightness of the combustion chamber, has to be done with a compression gauge. This gauge allows you to find the compression in the cylinder at the end of the compression stroke to compare it with the manufacturer's specification.

● How to do it

- ◆ Remove the injectors.
- ◆ Install the gauge in place of the injector.
  In case you pulled out just one injector to check only one cylinder, pull on the engine stop so it doesn't start on the other cylinders.
- ◆ Crank the engine for about five seconds. Read the gauge.
- ◆ Push on the gauge reset to bring the needle back to zero.
- ◆ Repeat the operation on the other cylinders.

● Readings

The readings in bars for each cylinder should be at least equal to the compression ratio. Compression ratio median reading:

- ◆ Direct injection: 24/1
- ◆ Indirect injection: 17/1 to 23/1

A weak reading across all of the cylinders means worn piston rings and cylinders. The difference between the cylinders should be below one bar. It shows that the cylinder with the low compression is faulty.

Low pressure on a single cylinder may come from a burned valve or faulty rings (broken rings) on that cylinder.

1 First loosen the leak vent-off pipe screw on the injector.

2 Save the copper seal washers.

3 Loosen the high pressure lines before loosening the injector plate.

4 Loosen the plate, completely remove the fuel line and remove the injector.

5 Install the compression gauge in place of the injector, tighten the plate.

6 Crank the engine. During this step it is best to get some help to activate the contact key. If you removed only one injector, pull on the engine stop so it doesn't start on the other cylinders.

   You will note during this step that the compression gauge needle rises in steps until it settles on the maximum reading, usually reached after 4 or 5 seconds. The needle will stay at that pressure. Reset the compression gauge to zero.

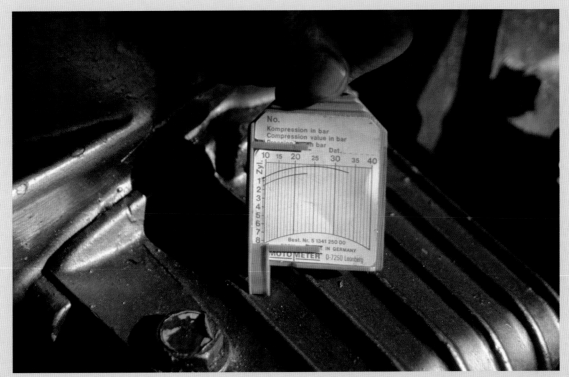

7 When you have readings for all of the cylinders (here, a two-cylinder engine) you can compare these with the manufacturer's specifications. The reading for each cylinder must be at least equal to the compression ratio.

- ◆ Direct injection: 27/7
- ◆ Indirect Injection: 17/1 to 23/1

Also, there shouldn't be more than one or two bars difference between the two cylinders. This indicates that the cylinder with the low compression is faulty. A weak reading across all of the cylinders means worn piston rings and cylinders. In the case shown here, the pressure in cylinder No 2 is clearly lower than that of cylinder No 1.

   To refine the diagnosis, at this stage it is possible to find out the condition of the rings and of the valves and their tightness. Remove the gauge and pour a cupful of new engine oil in the injector hole. Take the readings again. If you get a higher reading this time, it is because the rings are poor. The oil you poured in accumulated above the rings and created a seal. If you get a reading within one or two bars of the previous reading, it is because the valves are not tight; they are leaking. On the other hand, the rings and the pistons are in perfect condition. In this case, you have to remove the cylinder head and recondition the valve seats to re-establish the cylinder's tightness.

8 During reassembly, remember to position and hand tighten the high pressure lines on the injector holder; follow the torque tightening for the injector plate bolts; and to change the washer seals on the fuel return lines or to anneal them.

Checking the oil pressure

Simple

- 30 minutes
- Basic tools, oil pressure gauge

Four-stroke engines are equipped with journals (crank-shaft, camshaft) lubricated by pressurised oil. The life span of the engine depends on the oil pressure in the circuit. Reading the pressure tells us how worn the engine is. Every engine is equipped with an oil pressure warning light and/or an audible alarm (buzzer). When the light or the buzzer comes on, the engine is already suffering from lack of oil.

Test

- Get an oil pressure gauge that fits in place of the oil pressure switch.
- This test is done with the engine cold until the temperature rises to normal running temperature.
- Unscrew the oil pressure switch. Screw the flexible hose of the oil pressure gauge in its place.
- Start the engine.

Readings

We can never stress enough the importance of monitoring the oil pressure in the lubrication system.

Manufacturers fit engines with an oil pressure warning light that goes out when the pressure reaches 0.5 to 1 bBar/cm³. But since the correct pressure for an engine at cruising speed is about 4 bars, you have no precise indication of the real pressure in the lubrication circuit. Be careful. You think you're safe; your oil pressure light stays off. But inside your engine, the pressure may have stabilised at 1.5 bars. There can be several reasons for this: old oil or oil level too low, which means that the oil circulates too quickly, causing over-heating and loss of viscosity; the engine starts to wear. If you keep on motoring, you risk throwing a con-rod within a short time. This is serious because it means a total engine rebuild (replacing the bearings, machining the crank shaft etc)

If the pressure is 1 or 2 bars below specification, the engine is beginning to wear out. If the pressure is 2 bars below what it should be, there is a serious lubrication problem. If the pressure is 2 bars too low at any rev, and 1 bar too low at idle, it's time to investigate the problem without delay to avoid a serious and costly break down.

1 First find the oil pressure switch. Circular in shape, it is often located in the lower part of the engine block, not far from the oil filter.

2 Once the oil switch is removed, put the oil gauge in its place.

Diagram of an oil pressure alarm electrical circuit

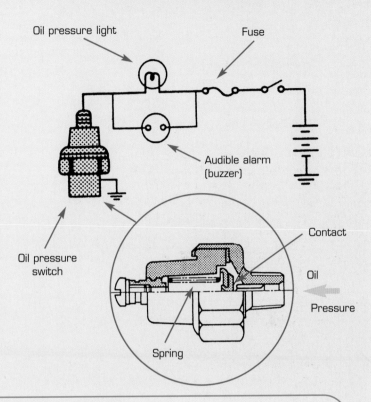

Oil pressure light

Fuse

Audible alarm (buzzer)

Oil pressure switch

Contact

Oil Pressure

Spring

3 Take this reading with the engine cold, then hot and with the engine in gear at every rev level.

Troubleshooting Table

Problems	What to check	How to check it	Solution
The light does not come on when the contact is 'ON'.	**1** Light burned out	**1** Visual inspection **2** The light does not come on even when the oil pressure switch is shorted	Replace the bulb
	2 Check oil pressure switch	The light comes on during the test	Replace oil pressure switch
The light stays on when the engine runs	**1** Oil level too low	Stop the engine and check oil level	Top up the oil
	2 Oil pressure too low	Check oil pressure	Check and repair the pump. Adjust the valve
	3 Faulty oil pressure	Faulty oil pressure switch	Replace the oil pressure switch
	4 Broken wire between light and oil pressure switch	Connect a separate wire between them	Repair the circuit

BREAKDOWNS

HOW DO YOU AVERT THEM? Track down the first signs of wear, any abnormal noises, unusual vibrations and leaks. Try to locate the cause of the problem. An engine rarely breaks down without a warning sign; there will be obvious symptoms. Learn how to recognise them. Also learn also how to tell which symptoms are not serious and those which are. The following tables will help you quickly locate the cause of a breakdown and direct you to the work sheet that will allow you to fix it.

How to use the tables

If a breakdown or a fault occurs, look for the description below that matches the fault observed.

Note the **associated numbers**. Then, starting with the first number shown, search for the probable cause of the breakdown and its solution on pages 217–19.

Problem description	Number to check (pages 217–19)
My engine:	
Doesn't start and emits black smoke	4, 14, 18, 11
Doesn't start and emits white smoke	10, 17, 11, 16
Doesn't start and doesn't smoke	1, 3, 6, 8, 9
Starts then stops	3, 5, 6
Has no power	3, 7, 9, 18, 19, 20
Misfires	6, 23, 3, 18, 25, 40
Doesn't reach its full revs	23, 28, 29, 7, 2, 36, 37, 38, 33, 34
Stalls when it is put in gear	27, 28, 23, 37, 38
Vibrates and knocks	63, 64, 35, 66, 65, 84, 85, 86
Overheats	44, 45, 46, 47, 48, 49, 50, 51, 52
Uses too much water	53, 54, 41, 16, 55
Emits white smoke	56, 16
Emits blue smoke	27, 58, 59, 60, 9, 61
Emits black smoke	5, 20, 2, 18, 11, 24, 9, 62, 33, 38, 37, 31, 34
Uses too much fuel	5, 19, 18, 11, 24, 9, 20, 22, 33, 34
Uses too much oil	73, 27, 4, 74
The oil pressure is low	28, 67, 68, 61, 69, 70, 71, 72
The engine vibrates when in gear	63, 65, 64, 66, 85, 86, 84
The charge light stays on	75, 76, 77, 78
The starter doesn't start the engine	13, 12, 80, 79
The boat doesn't move under power	81, 87, 82, 88, 83

	Probable causes	Solutions
1	Empty fuel tank.	Fill the tank and bleed the fuel lines.
2	Bad fuel quality.	Drain the tank, use a fuel recommended by the manufacturer.
3	Clogged fuel filter.	Change the filters, bleed the fuel lines.
4	Damaged air filter.	Replace the air filter (see manual).
5	Clogged air filter.	Change the air filter cartridge (see manual).
6	Air in the fuel circuit.	Check for leaks, bleed the system.
7	Throttle control badly adjusted.	Adjust the control cable.
8	The engine stop stays on.	Check the ignition circuit and the solenoid. Possibly check the injection pump.
9	Insufficient compression.	Check the condition of the valves, rings, head gasket.
10	Faulty pre-heating.	Check the electrical circuit, check the glow plugs.
11	Pump adjustment.	Check and adjust the injection pump.
12	Battery not sufficiently charged.	Recharge the battery.
13	Battery terminals loose or dirty.	Clean and tighten battery terminals.
14	Insufficient revs when cranking.	Check the battery, electrical circuit, engine oil quality.
15	Faulty starter motor.	Remove, test the starter motor.
16	Blown head gasket.	Change the head gasket, check the cooling system, change the oil if cooling water has got into it.
17	Throttle doesn't open when cranking.	Check the throttle controls, adjust if needed.
18	Faulty injector.	Remove the injector, adjust or change if needed.
19	Injector pump not adjusted	Get the pump checked at an authorised centre.
20	Partly clogged exhaust.	Check the exhaust system.
21	Engine running temperature too high.	Check the cooling system.
22	Engine temperature too low.	Check the cooling system (thermostat).
23	Idle too slow.	Adjust the idle.
24	Valves not adjusted.	Adjust the valves.
25	Piston stuck or clogged rings.	Check the compressions, restart the engine.
26	Faulty injector pump.	Get the pump checked at an authorised centre.
27	Oil level too high.	Adjust the level.
28	Unsuitable oil viscosity.	Check; use an oil with manufacturer's recommended viscosity.
29	Bad engine room ventilation.	Check and re-establish the engine room ventilation.
30	Dirty propeller.	Clean the propeller.
31	Propeller too big.	Use a suitable propeller.

Probable causes		Solutions
32	Wrong reduction ratio.	Use a suitable reduction ratio.
33	Dirty hull.	Clean the hull.
34	Overloaded boat.	Unload, get the boat back on its waterline.
35	Damaged propeller.	Remove, change or repair the propeller.
36	Clogged fuel tank breather.	Check the breather.
37	Stuffing box too tight.	Adjust the stuffing box.
38	Line caught in the propeller.	Clear the propeller; check the shaft alignment.
39	Stop lever open or not adjusted.	Close or adjust the decompression/stop lever.
40	Leaking valves. (valves burnt or damaged seats).	Remove and recondition the engine head.
41	Leaking head gasket.	Remove the head, change the gasket, check the cooling system.
42	Worn rings. (broken or clogged rings, a fair bit of smoke exits from the breather or the oil filler cap.)	Recondition the engine.
43	Thrown rod.	Recondition the engine.
44	Closed seacock.	Open seacock.
45	Clogged salt water filter.	Clean the filter.
46	Low sea water flow.	Check the cooling system (pump, thermostat...).
47	Pinched or pierced hose.	Change the hoses.
48	Loose or broken water pump belt.	Tighten or change the belt.
49	Water passages clogged.	Clean the block, descale the water passages.
50	Clogged heater exchanger.	Remove, clean the heat exchanger.
51	Faulty water pump	Change the water pump.
52	Faulty thermostat.	Check, change the thermostat.
53	Faulty heat exchanger or expansion bowl cap.	Check the cap calibration spring.
54	Various leaking water hoses.	Check the cooling system for leaks.
55	Leaks in the heat exchanger.	Remove heat exchanger, get it checked under pressure.
56	Cooling water vaporises in the exhaust manifold or muffler.	Check the water circuit (scale).
57	Timing too fast.	Check, adjust the injection pump timing.
58	Worn valve guides.	Change the guides, retighten the head.
59	Incomplete combustion, timing too slow.	Adjust the injection timing.
60	Worn cylinders and rings.	Recondition the engine.
61	Oil pressure regulating valve.	Check the oil pressure, adjust if necessary.

	Probable causes	Solutions
62	Too much fuel.	Check fuel flow, the maximum flow stop and the regulator. Consult an injection specialist.
63	Loose engine block.	Tighten the engine block, check the shaft alignment.
64	Play at the rear bearing bracket.	Check the play, check the bearing bracket fixings.
65	Play at the rear of the propeller shaft. (worn cutless bearing)	Change the cutless bearing.
66	Bent, dirty propeller (full of barnacles).	Remove, clean and get a specialist to check it.
67	Low oil level in the sump.	Top up the oil.
68	Faulty oil pressure switch.	Replace the switch.
69	Clogged oil filter.	Change the filter.
70	Oil strainer clogged.	Remove, clean the strainer.
71	Worn oil pump.	Replace the oil pump.
72	Too much play at the con-rod bearings and crank shaft bearings.	Recondition the engine.
73	Oil leaks.	Check for leaks, change the seals, check the seal seats.
74	Worn out engine (valve guides, rings).	Recondition the engine.
75	Faulty connections.	Reconnect.
76	Broken alternator belt.	Replace the belt.
77	Faulty regulator.	Check the voltage. Change the regulator if necessary.
78	Faulty alternator.	Remove, test the alternator.
79	Stuck or worn brushes.	Check the starter motor.
80	Bad ground connections.	Check the starter circuit, particularly the starter ground.
81	Shaft couplings loose, loose or shorn grub screws or pin.	Tighten the couplings, check the grub screw or pin.
82	Sheered propeller pin.	Remove the propeller, replace the pin, check the propeller shaft alignment.
83	Loss of propeller.	Replace the propeller.
84	Bent propeller shaft.	Check the shaft, change the propeller shaft, adjust alignment of engine/propeller shaft.
85	Loose propeller shaft anode.	Tighten the anode.
86	Locked folding propeller; One blade stays folded.	Clean the propeller, check that it opens freely.
87	Faulty gear controls adjustment.	Adjust the controls.
88	The gears slip.	Check the gear box (change the cones or disks)

WINTERISING

DEALING WITH A BREAKDOWN will be necessary from time to time, but it is better to prevent it in the first place. This of course means regular maintenance but also complete winterising of the engine and its accessories. In fact, it is during its inactivity that your engine can suffer the most, if only because it is in a hostile marine environment. This process will take you a day's work laying up and half a day before the season starts but will get you off to a good start on the water.

WHY DO YOU NEED TO WINTERISE?

As the end of the sailing season approaches, it is time to think about winterising your engine to protect it from corrosion. This task is within everyone's ability, especially for small and medium sized engines. However, if you feel that it is beyond your capabilities, don't hesitate to entrust it to a professional service engineer because corrosion acts with surprising speed. In fact, the majority of mechanical failures are often the result of poor maintenance and, in many cases, of bad winterisation.

During the period of inactivity during the winter, humidity levels rise causing attack by the salty air, leading to corrosion. This will make the hull and engine vulnerable to all sorts of breakdown. The winterisation steps require only a few tools but lots of care and meticulousness. It needs to be done within 15 days of the last use. If you aren't sure you can carry out this task, especially if you have a large engine, don't hesitate to call in the professionals.

How to do it

The winterising procedure is more or less the same for any engine. If you don't winterise your engine, and the boat stays in the water, the engine must be run at least every 15 days to avoid corrosion. If possible, let it run in gear to increase the engine temperature. If the boat isn't going to be used for over a month, it should be protected in the same way as for winter lay up.

Winter storage must be well planned. Start by doing a test run and carefully check all the engine functions. Note any faults. If something needs to be repaired professionally, make an appointment as soon as possible so the job can be done during winter when the service engineers are less busy. Winterising is preferably carried out, as explained later, with the boat still in the water.

TWENTY STEPS TO GOOD WINTERISING

1 Warm up the engine, then check the fuel, lubrication and cooling lines. Eliminate any leaks.
2 Stop the engine. Drain the oil sump and change the oil filter.
3 Fill the sump with new oil (minimum quality CD, API standard).
4 Start the engine, check the oil pressure and make sure there are no leaks at the filter.
5 Stop the engine and top up the oil if needed.
6 Replace the fuel filters to make sure there's no water or dirt in the circuit.
7 Fill up the fuel tank and plug the breather vent.
8 Close the seacock.
9 Drain and flush the seawater circuit.
10 Fill the circuit with a rust inhibiting liquid (radiator coolant)

   If the engine is equipped with an anti-siphon valve, it has to be removed and cleaned.

Every boatowner should have a basic tool kit aboard.

Draining the oil requires only limited equipment. The drill is practically the same on any engine.
   A common rule: The oil has to be drained when the engine is at running temperature.

Flush the seawater circuit. Disconnect the saltwater hose at the seacock. Fill a bucket with radiator coolant.
   When the liquid begins to come out of the exhaust, stop the engine. Your engine is now protected.

Change all the fuel filters, then bleed the circuit. Remember to top up your fuel tank; it will prevent condensation problems.

Long term winterising

Remove the injectors and spray anti-corrosion oil into the injector holes. If you don't have that type of oil, plain engine oil does the job to a certain extent.

Slowly turn over the engine (one turn), then replace the injectors with new washer seals. Remove the impeller from the water pump. If it is intact, keep it in a plastic bag. This will keep the blades from deforming.

Remove the exhaust pipes and spray anti-corrosion oil in the exhaust manifold. Close it tightly with adhesive tape.

For engines with an overhead camshaft, you can also unbolt the shaft or each individual rocker arm to isolate the cylinder chambers.

Cleaning the anti-siphon valve

1 It is sometimes difficult to reach the valve because it has to be placed well above the water line (minimum 40cm).

2 After removing the valve, check it for clogging. Salt crystals or other dirt may prevent it from opening and closing.

Cleaning the salt water filter

1 The water filter is an indispensable safety device. Being above the waterline, it can be easily cleaned.

2 When you motor through seaweed, the filter does its job. The clear cover allows inspection without opening it.

3 Greasing the seal with Vaseline will make it easier to dismantle the next time it has to be cleaned.

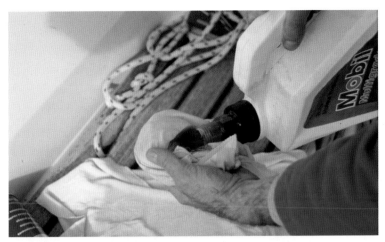

Wet a rag with oil to 'seal' it.

11 Remove the oil filter or the inlet tube and spray anti-corrosion oil in the inlet manifold. Plug the manifold with an oil-soaked rag or airtight adhesive tape.

Plug the air filter to isolate the cylinder chamber. But remember to remove the plug before starting the engine.

12 Plug the exhaust outlet with adhesive tape or an oily rag.

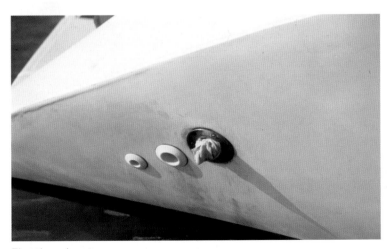

Plug the exhaust.

13 Protect the electrical circuit. The circuit wiring has a lot of connections where humidity can penetrate. A water repellent spray will efficiently protect them and will prevent electrical problems.

14 Disconnect the battery, clean the terminals. Batteries can be stored on board if they are fully charged. If in doubt about their charge state, remove them so they can be trickle charged and stored where they cannot freeze.

The connections and various components of the electrical circuit are very sensitive to humidity.

Spraying with WD40 effectively protects your engine from corrosion.

Disconnect the connections and spray with WD40 or equivalent.

15 Slacken the belt on the alternator and water pump.

16 Clean the engine compartment, especially the area under the engine, to remove anything that can create odours and to reduce humidity.

17 Inspect and clean the engine. Touch up the paint to prevent corrosion.

18 Spray the entire engine with a water repellant spray such as WD40.

19 Oil the throttle, the gear controls, the stop pull knob and the contact key.

20 If the engine is fitted with a traditional drive shaft, tighten the stuffing box nut to eliminate any water drips.

In the case of a rotary seal, there's no need for any particular precaution, other than checking the condition of the spring seal and the clamps. If it is a lip seal, apply water-resistant grease.

Slacken the belts to extend their lifespan and save the bearings from stress marks. Check their condition; change if necessary.

Remember to oil the stop pull knob.

If the protective steps described are followed, your engine shouldn't suffer from deterioration due to corrosion during the lay up period. Even so, such a job list can't be exhaustive; each engine has different requirements and each owner has a different programme and different ability and experience. Therefore this list is a general guide that needs to be completed or adapted according to your personal circumstances.

Check that the engine controls are functioning well.

Finally, the long awaited sailing season is here. You are back on your boat again after the winter break, but before casting off the lines, you have to de-winterise. A few hours' work and your engine will be able to start in the best possible condition.

Long term winterising

Tighten the camshaft bolts and adjust the rocker arms.

Re-install the pump impeller.

◆ Unplug the air inlet manifold, the exhaust, and the fuel tank breather.
◆ Reinstall the battery.
◆ Check the water level in the heat exchanger, if your engine has one.
◆ Open the seacock.
◆ Adjust the tension on the belt that drives the engine accessories.
◆ If you have a traditional stuffing box, loosen it. The propeller shaft should spin by hand, with the box out of gear.

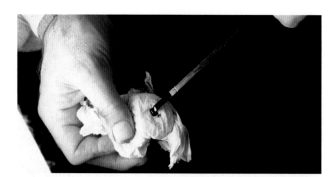

Have a look at the oil level (above left), at the seacock and the switches (left and above).

- Move the throttle and the gear controls.
- Check the oil level once more.
- Turn on the switches.
- Crank the engine while decompressing to raise the oil pressure.
- Your engine is now ready to start.

Starting the engine

You might find the engine a bit reluctant to start after winterising so follow this procedure:.

- Move the engine stop control to its start position.
- Turn on the engine room blower (if there is one) and let it run for a few minutes before starting the engine.
- Make sure the fuel valve and seacock are open.
- Put the throttle out of gear and rev it half way. On some engines this lever should be wide open; on others, it needs to be on idle. Check the engine manual to avoid making a mistake.
- Turn on the contact. The oil pressure light and buzzer and charge light should come on and the different gauges shouldn't show any irregular readings.
- Pre-heat if needed.
- Turn the key to 'ON' or press the start button and keep it on until the engine starts.
- After a few seconds the engine should begin to cough, then settle on a regular tick-over. If this isn't the case and you have difficulty starting it, don't persist. Get professional help.
- Immediately after starting, check that the indicator lights on the control panel are off. Stop the engine if a buzzer or a light stays on. Set the throttle at a moderate speed, ie fast idle 1000rpm–1500 rpm.
- Check that cooling water spits out of the exhaust. If it doesn't, stop the engine and look for the cause.
- Check that there are no leaks in the water circuit and no fuel leaks (around the filters).
- At the control panel, check the water temperature, the engine temperature and the oil pressure, etc.
- With the engine hot, check the idle speed.
- Engage forward gear then reverse and check that the gear isn't slipping. The engine purrs and now you're ready to get going again for another season.

Activate the key or button to turn the contact on. The oil and charge lights must come on. Pre-heat, then activate the starter. The engine ought to start after a few seconds and all the lights should go out.

Check that the water pump is working properly by watching the flow from the exhaust.

INDEX

DATE	TASK CARRIED OUT	

RECORD

COMMENTS	ENGINE HOURS

MAINTENANCE

DATE	TASK CARRIED OUT	

RECORD

COMMENTS	ENGINE HOURS